THE COLONIALITY OF MODERN TASTE

This book analyzes the coloniality of the concept of taste that gastronomy constructed and normalized as modern. It shows how gastronomy's engagement with rationalist and aesthetic thought, and with colonial and capitalist structures, led to the desensualization, bureaucratization and racialization of its conceptualization of taste.

The Coloniality of Modern Taste provides an understanding of gastronomy that moves away from the usual celebratory approach. Through a discussion of nineteenth-century gastronomic publications, this book illustrates how the gastronomic notion of taste was shaped by a number of specifically modern constraints. It compares the gastronomic approach to taste to conceptualizations of taste that emerged in other geographical and philosophical contexts to illustrate that the gastronomic approach stands out as particularly bereft of affect. The book argues that the understanding of taste constructed by gastronomic texts continues to burden the affective experience of taste, while encouraging patterns of food consumption that rely on an exploitative and unsustainable global food system.

This book will appeal to students and scholars interested in cultural studies, decoloniality, affect theory, sensory studies, gastronomy and food studies.

Zilkia Janer is a Professor of Global Studies and Geography at Hofstra University in Hempstead, New York.

Routledge Research on Decoloniality and New Postcolonialisms

Series Editor: Mark Jackson, Senior Lecturer in Postcolonial Geographies, School of Geographical Sciences, University of Bristol, UK.

Routledge Research on Decoloniality and New Postcolonialisms is a forum for original, critical research into the histories, legacies, and life-worlds of modern colonialism, postcolonialism, and contemporary coloniality. It analyses efforts to decolonise dominant and damaging forms of thinking and practice, and identifies, from around the world, diverse perspectives that encourage living and flourishing differently. Once the purview of a postcolonial studies informed by the cultural turn's important focus on identity, language, text and representation, today's resurgent critiques of coloniality are also increasingly informed, across the humanities and social sciences, by a host of new influences and continuing insights for different futures: indigeneity, critical race theory, relational ecologies, critical semiotics, posthumanisms, ontology, affect, feminist standpoints, creative methodologies, post-development, critical pedagogies, intercultural activisms, place-based knowledges, and much else. The series welcomes a range of contributions from socially engaged intellectuals, theoretical scholars, empirical analysts, and critical practitioners whose work attends, and commits, to newly rigorous analyses of alternative proposals for understanding life and living well on our increasingly damaged earth.

This series is aimed at upper-level undergraduates, research students and academics, appealing to scholars from a range of academic fields including human geography, sociology, politics and broader interdisciplinary fields of social sciences, arts and humanities.

Transdisciplinary Thinking from the Global South
Whose problems, whose solutions?
Edited by Juan Carlos Finck Carrales and Julia Suárez-Krabbe

The Coloniality of Modern Taste
A Critique of Gastronomic Thought
Zilkia Janer

For more information about this series, please visit: https://www.routledge.com/Routledge-Research-on-Decoloniality-and-New-Postcolonialisms/book-series/RRNP

THE COLONIALITY OF MODERN TASTE

A Critique of Gastronomic Thought

Zilkia Janer

Routledge
Taylor & Francis Group

LONDON AND NEW YORK

Cover image: Gargantua's Meal by Gustave Doré. Photo courtesy of Bibliothèque nationale de France.

First published 2023
by Routledge
4 Park Square, Milton Park, Abingdon, Oxon OX14 4RN

and by Routledge
605 Third Avenue, New York, NY 10158

Routledge is an imprint of the Taylor & Francis Group, an informa business

British Library Cataloguing-in-Publication Data
A catalogue record for this book is available from the British Library

ISBN: 978-1-032-36403-2 (hbk)
ISBN: 978-1-032-36417-9 (pbk)
ISBN: 978-1-003-33183-4 (ebk)

DOI: 10.4324/9781003331834

Typeset in Bembo
by SPi Technologies India Pvt Ltd (Straive)

Dedicado a Sanjib

CONTENTS

ACKNOWLEDGMENTS

When I began reading the classics of nineteenth-century gastronomy, I was not planning to write a book on the subject. However, I did not find in gastronomic texts the sensuous celebration of the pleasures of the table for which they are known. What I found was much more ambiguous and problematic. This book is the result of my efforts to explain the discrepancy between the reputation of gastronomic texts and what I found in them.

This book is the culmination of many years of work across disciplines and continents. I would not have been able to write it without the help of many generous people who took an interest in my project. My gratitude to all of them goes beyond what I can express in a few lines.

I am thankful for the feedback given by co-panelists and participants at conferences organized by the Association for the Study of Food and Society, the Cultural Studies Association, the Latin American Studies Association and the Asian Borderlands Research Network, where I presented preliminary versions of many of the arguments developed in this book. In particular, Krishnendu Ray, Ken Albala, Amy Reddinger and Lena Burgos-Lafuente gave engaging responses to my talks. I am also grateful to Nandana Dutta, Xonzoi Barbora, Siddiq Wahid and Preminder (Pami) Singh for inviting me to give lectures at Gauhati University, the Guwahati campus of the Tata Institute of Social Sciences, the Awantipora campus of the Islamic University of Science and Technology and the India International Centre, respectively. The insightful comments of the audience of these lectures kept coming to my mind while writing the manuscript.

I am grateful to Tejbir Singh, editor of *Seminar*, and to Kanak Dixit, founding editor of *Himal Southasian*, for publishing my earliest thoughts on the geopolitics of taste. I thank Prasenjit Duara, Anne Feldhaus and Andrea Lorene Gutiérrez for generously taking the time to respond to queries and offering productive research leads. I am also grateful to Walter Mignolo for being a constant source of inspiration and encouragement.

The exchange of ideas that helped me develop my argument took place in kitchens and at table as much as at libraries and academic conferences. The Janer family in Puerto Rico and the Baruah family in Assam, India, have been responsible for my life-long sensory education. I cannot thank them enough. I will always be grateful to those who have further expanded the range and depth of my gustatory understanding: Sabyasachi (Shobo) Bhattacharya, Khin Khin Khant, Dolly Kikon, Amrith Lal, Jacob Mathew, Mrinal Miri, Sujata Miri, A.S. Panneerselvan, Kalpana Raina, Sabita Radhakrishna, Arshiya Sethi, Shahid Siddiqui, Roongfa Sringam and Siddharth Varadarajan. While each one of them helped me appreciate specific taste cultures, collectively they taught me to experience food in a more fully sensorial way.

Leaves of absence and a series of modest grants from Hofstra University allowed me to conduct research at the National Library of France, the British Library and the New York Public Library. I am indebted to my colleagues in the department of Global Studies and Geography for their support, as well as to Pepa Anastasio, Suzanne Berman, Chandler Carter, Brenda Elsey, Vicente Lledó-Guillem and Kathleen Wallace for offering advice on their respective areas of expertise. Finally, I thank Sanjib Baruah for our daily feasts of stimulating conversation.

INTRODUCTION

What Is Gastronomy?

Gastronomy is a discourse that defined a distinctly modern approach to the sense of taste. This would be a sufficient definition if it were not for the mystifications that surround the idea of modernity. Modernity usually evokes a narrative of constant human progress led by the West from the Enlightenment onwards. Following this narrative, gastronomy has been celebrated as a breakthrough in the human cultivation of the sense of taste. Gastronomy, particularly as formalized by mostly French writers in the nineteenth century, is supposed to be a high point in the history of taste cultures. Such a triumphalist stance is the result of approaching gastronomy from the perspective of the modern/colonial discursive field in which it arose. However, the unqualified celebration of modern culture has become untenable in the light of the many substantial critiques that have multiplied in the past few decades. These critiques come from a variety of perspectives, ranging from feminism and post-modernism to postcolonialism and decoloniality. Whereas these critical perspectives focus on different issues, they all share a sense that the narrative of modernity is a harmful fairy tale. Such critiques of modernity, however, have had insufficient impact on the scholarship of gastronomy. While gastronomy was indeed defined by modernity, this should not be taken as proof of advancement or superiority. On the contrary, when looked at from a critical global historical perspective that questions the narrative of modernity and rejects epistemic Eurocentrism, gastronomy is revealed as a troubling taste culture that we all need to seriously interrogate and overcome.

Taste, like all the senses, never exists in a natural state.[1] There are different taste cultures because the sense of taste is constructed in different ways in different times and places. A taste culture is not only a set of ways of producing, preparing, serving and consuming food but also a way of thinking about taste that shapes how taste can be experienced. All taste cultures both curb and enhance the experience of taste,

DOI: 10.4324/9781003331834-1

while being attentive to meaning and pleasure. The meaning of taste comes from how people in each geohistorical location relate it to their larger systems of thought, particularly their notions of subjectivity, knowledge, beauty, health, morality and transcendence. A taste culture is thus a set of food-related practices informed by systems of thought. All human societies develop taste cultures, and many have had the inclination and resources to create taste cultures of remarkable sophistication. Historical records offer evidence of the great diversity of sophisticated taste cultures in every region of the world before, during and after the emergence of gastronomy. Gastronomy is not the first time that taste was valorized or elevated, it is only the first time it was valorized according to distinctively modern ideas. If we do not take modernity as the necessary destiny of humanity, this means that gastronomy is just one contingent culture of taste among many. Gastronomy's strong and lasting impact comes not from excellence or superiority but from its articulation with modern structures of power.

Modernity is above all a narrative that has constructed the idea of the West and celebrated its achievements without acknowledging that they were made possible by the sustained appropriation and exploitation of the resources, labor and knowledge of the peoples and places that have been colonized and otherwise subjected by the gradual establishment of global capitalism.[2] Modernity is not something that European peoples developed all by themselves and that has slowly spread throughout the world, as the narrative of modernity would have it. Instead, modernity is the culture that Europe developed in the process of establishing and holding on to the central position that it acquired by the colonization of the Americas,[3] and which continued through the colonization and exploitation of other vast regions of the world. As Walter D. Mignolo put it, "[M]odernity/coloniality are two sides of the same coin."[4] Because the narrative of modernity hides or downplays coloniality, a critical understanding of modernity must confront how coloniality constituted the modern. In this light, the identification of gastronomy with modernity should not be taken as something to celebrate unreservedly. As a modern discourse, gastronomy was defined by the inequalities of global capitalism and colonialism, and it was instrumental in the modern construction of race. The point is not to negate that modernity achieved unprecedented food abundance and variety for the most powerful countries but to face the fact that this cornucopia was achieved and continues to be sustained at the expense of most of the world.

The discourse of gastronomy has at its core a narrative that defined not only gastronomy itself but all other taste cultures as well. This narrative, which continues to be rehearsed in countless popular and scholarly accounts, inflates the significance of the distinguishing characteristics of gastronomy and undervalues the taste cultures of other times and places. The distortion of the comparative appreciation of different taste cultures takes off from the way in which modern thinkers systematically portrayed themselves as bearers of a culture that has a series of characteristics that put it in a superior class apart from all others. The modern distinction between modern and traditional cultures put modern cultures on a pinnacle that traditional cultures had yet to attain. The specific characteristics and trajectories of all other cultures became obscured under their blanket classification as "traditional."[5] Following the pattern of modern thinkers and ideologues, the narrative of gastronomy has established it as one more expression of the supposed cultural superiority of the West. Not surprisingly, the characteristics that are usually invoked to establish the exceptionality and superiority of gastronomy come straight from the list of characteristics that are said to distinguish modern from traditional cultures. One of the most celebrated characteristics of gastronomy, which is usually invoked to establish it as a more evolved taste culture, is its autonomy from direct external constraints coming from other fields, like for example dietetics.[6] However, to understand gastronomy without the distortions of the narrative of modernity, we must be willing to let go of the myth of gastronomic autonomy.

The Myth of the Liberation of Taste

The myth of the liberation of taste established that the development of gastronomy in nineteenth-century France marks the liberation of taste from non-gustatory influences. Priscilla Parkhurst Ferguson defined modern cuisines as characterized by freedom:

> If place grounds traditional cuisines, freedom from that same place liberates modern cuisine – freedom to experiment and recast the material, freedom from the community, freedom to make cuisine an intellectual and aesthetic as well as a material and sensual experience.[7]

Whereas it is widely recognized that rationalism and aesthetics came to define modern taste, this is not usually regarded as interfering with its stated freedom. Instead, rationalism and aesthetics are said to have elevated and intellectualized taste. While it is fair to say that gastronomy downplayed the influence of previous religious and medical systems, the myth of the liberation of taste underestimates the severity of the limitations that rationalism and aesthetics imposed on gastronomy and the experience of taste. Gastronomy indeed represents a dramatic change in the way that Europeans thought about taste, but this change by no means could be characterized as a liberation. Gastronomy perhaps seems more autonomous than other cultures of taste because it was given a name and its principles were explicitly established in print. However, even a relative gastronomic autonomy of the kind that Pierre Bourdieu argued characterizes the cultural field was more an aspiration than a reality because gastronomy was fundamentally shaped and ruled by laws that were not of its own making.[8] The rules that shaped gastronomy came not only from its subordination to established fields with a relative autonomy like philosophy and science but also from the logic of globalizing capitalism and imperialism. Gastronomy was the result of the reconfiguration of European taste culture in the context of an emerging modern culture and the changing fields of power, both internally and globally. Gastronomy represents the shift from an aristocratic taste culture to one defined by bourgeois subjectivity, the capitalist market and Europe's sense of cultural superiority in the imperial era. Gastronomy did not and could not possibly liberate taste from non-gustatory influences, given that taste is the result of an experience shaped by the sociocultural context in which it unfolds. However, the fact that gastronomy has been praised as having achieved the liberation of taste reveals that the modern ideal of taste is one in which taste is unaffected by its complex interactions with psychosocial and geohistorical variables.

The myth of the liberation of taste stands in the way of a nuanced understanding of the character of the significant transformations in the approach of modern societies to eating and the sense of taste. This myth misses the same crucial point that Michel Foucault pointed out was missing in the understanding of the attitude of modern societies regarding sex. Foucault explained that what characterizes modern Western societies is not the repression of sex, as it is generally believed, but its placement into discourse.[9] In his view, the "repressive hypothesis" obscures the specific ways in which the approach to sex changed after the eighteenth century, when the religious framework for sexual rules of behavior partly disappears and new medical and juridical frameworks compete to take its place.[10] I argue that the myth of the liberation of taste constitutes an inverse but similarly flawed liberation hypothesis. What happened to taste in Europe after the Enlightenment was not its liberation but its placement into a new configuration of old and new discourses in the shape of gastronomy as a modern discipline of taste. The Foucauldian concept of the discipline is a form of social control that regulates the behaviors of individuals. In this book, I propose that gastronomy is a discipline that regulates the relationship between eaters and food and between eaters by limiting the affective power of taste.

There is no such thing as an autonomous culture of taste. In all taste cultures, there is an interplay of different systems of thought that elevate, socialize, enhance and curb the act of eating. Different expressions of what modernity calls religion, medicine, rationalism and aesthetics have always been related to each other in all taste cultures, taking specific configurations in each geohistorical location. It is too simplistic to think that rational and aesthetic concerns were absent from the taste culture of so-called traditional societies and that religion and medicine had little role in shaping modern gastronomy. It is also erroneous to presuppose that religion and medicine can only hamper the enjoyment of taste and that rationalism and aesthetics can only enhance it. Religion and medicine in many cases provide intellectual and aesthetic structures that can enhance gustatory enjoyment, while some secular intellectual and aesthetic structures can stifle such enjoyment. Instead of uncritically celebrating gastronomy, the focus of this book is to explore and understand how the contingent modern articulation of religious, medical, epistemological, aesthetic and imperial discourses shaped how taste could be thought about and experienced in modern societies. I reject the presupposition that the specific concept of taste articulated in gastronomy is more desirable by virtue of being consistent with modern ideas and globally spread. Many of the refined taste cultures that the discourse of gastronomy claimed to have superseded actively encouraged the enjoyment of taste by engaging with the affective aspects that gastronomy in many ways aimed to suppress. When evaluating different taste cultures, we should not assess how closely they follow modern ideals and prescriptions, as if they were a universally desirable point of arrival. What should be assessed is how rich and full is the experience of taste that they enable. Put in this perspective, gastronomy falls unequivocally on the more restrictive end of the spectrum of the diversity of taste cultures across space and time.

The consolidation of gastronomy in the nineteenth century was the cumulative result of many transformations that began in the early modern period. These transformations might seem like a liberation in the European context, given that until then the general cultural attitude toward the sense of taste in Europe was overwhelmingly negative. Christian thought enforced a disinterested attitude toward food and its pleasures, so people interested in taste cultivation always risked falling into the cardinal sin of gluttony. In contrast, other major religions did not antagonize and even encouraged the pleasures of taste. Jewish and Islamic thinkers see food as a divine gift to be enjoyed. Koranic passages have served to justify culinary pleasures, and they have been used in the preambles of classic cookery books.[11] In Indic thought, cooking is a fundamental metaphor,[12] and many Hindu temples are associated with specific foods and delicacies. Whereas in other religions cooking and eating did not lack intellectual and aesthetic legitimacy, and there was the possibility of excelling in cooking and enjoying food with pleasure rather than shame, Christian thought mostly encouraged temperance if not asceticism. None of these systems of thought is monolithic, and there are ascetic and sybaritic strains in all of them. But while the emergence of gastronomy is generally recognized as a turning point that opened the possibility of exploring the pleasures of taste in the Christian European context, it should be noted that this possibility had always been open to different degrees elsewhere. Gastronomy then is not a turning point in global history. It is only a turning point in the local European context, which brought Europe closer to other parts of the world, where positive valorizations of taste had been ongoing for centuries.

Western philosophy has a long history of hostility toward the pleasures of taste and toward the body in general. From classical antiquity, Western philosophers shunned taste as a lower sense that was morally suspect and incompatible with reason.[13] Up to the early modern period, all kinds of texts, whether religious, medical, moral, cultural, social or symbolic, consistently devalued taste.[14] In literature, food was mentioned only in lower genres like comedy and satire. The pleasures of taste were excluded from written texts and polite speech until the mid-seventeenth century, when a new culinary sensibility emerged and cookbooks began to include prefaces that openly discussed taste in an appreciative manner.[15] However, the transformation of the attitude toward taste in modern Europe was not one from repression to freedom. It was a transformation from a culture in which taste was considered a liability best left unspoken to one in which it could at least be spoken and written about. The condemnation and anxiety regarding taste as a gateway to pleasure and sin did not disappear, but it became possible to talk about taste in an approving way as long as certain rules of acceptability were followed. These rules involved either invoking taste in a figurative manner or presenting the interest in taste as important for physical or social health. The plain and open celebration of gustatory pleasure for its own sake remained objectionable.

The first figurative use of the word "taste" in Europe emerged in sixteenth- and seventeenth-century mystical literature, thanks to the influence of authors like the Spanish Teresa of Avila and John of the Cross.[16] It is worth noting that these authors were influenced by Sufi and Jewish mystics, which indicates that the figurative use of taste that seemed revolutionary in Europe was already well established in other traditions of religious thought that did not condemn sensual pleasures. The figurative use of the word "taste" in European mystical writings endowed taste with spiritual value but at the expense of turning it into a non-corporeal sense. In order to taste the divine, you had to put down physical gustatory taste. The figurative use of taste was further developed in the eighteenth century with the emergence of aesthetics. Modern aesthetic philosophy is a thinly secularized version of mystical discourse in which physical taste became only a metaphor for the appreciation of the beautiful, conceived as universal and disinterested in direct opposition to physical pleasure. In both mysticism and aesthetics, the relative elevation of the standing of taste occurred through the negation of the body. In contrast, thought traditions that did not separate the mind from the body had always considered taste as an object of thought.

Cooks were the first ones in Europe to write about taste without wishing the body away. In the eighteenth century, cooks took it upon themselves to discuss taste and to make it acceptable by claiming the importance of cuisine for physical and moral health. New styles of cooking were developed to conform to shifting medical theories on health and digestion. The moral structures of sin and guilt that guided the approach to food and taste in earlier periods did not disappear, they were transcribed into the corporeal domain in secular language.[17] While attempting to lift the remarkably low status of taste in Europe, cooks could rarely invoke taste as pleasure without medical and ethical justifications. The pleasures of food and taste had to be presented as moderated by reason. Faced with this, the discourse of early modern cooks ended up advocating a notion of taste that took into account the body but did not engage with bodily enjoyment. They were preoccupied with preserving the health of the body rather than in providing pleasure.

From having no place in lettered and polite discourse to having a subordinated place in mystical, aesthetic and culinary discourses, taste acquired a discourse and a written genre of its own. Gastronomy, literally "rules of the stomach," was a new disciplinary knowledge that took its most notable shape in the book *The Physiology of Taste* written by French author Jean Anthelme Brillat-Savarin in 1825.[18] Although Brillat-Savarin is usually credited with inventing the gastronomic writing genre, his contribution was not so much inventing the genre as writing down in one volume the gastronomic ideas that had already gelled in the early nineteenth century. According to E.C. Spary, by the time that Brillat-Savarin wrote, it was no longer necessary to negotiate the opposition between learned sobriety and gastronomic self-indulgence.[19] Submitting to the parameters of modern rationalism and aesthetics, gastronomy created a notion of taste that obeyed supposedly universal rules.

This notion of taste downplayed any aspects that exceeded the ideal of rational objectivity. Gastronomy conceived of taste as an objective quality of foods and reduced the role of the eater to grasping this singular objective taste with as much accuracy as possible. Variations in perception that differed from the purported gastronomic universal were stigmatized and racialized. I argue that the creation of a separate discursive field for taste was more the result of the marginalization of taste in Europe than a sign of even relative autonomy. One of the main concerns of gastronomic writing was to demand recognition by the more powerful rational and aesthetic philosophical discourses by mimicking them and fashioning a concept of taste that would no longer be shunned by them. Gastronomy was more concerned with gaining intellectual acceptability for the interest in food than with exploring taste on its own terms.

Foucault explained that modern discourses constituted sex as a matter of truth,[20] and I would suggest that gastronomy did the same with taste. At one level, the truth that gastronomy wanted to reveal was the truth of taste itself. Believing that foods contained a single objective taste waiting to be grasped by knowledgeable palates, new cooking techniques were developed to showcase their supposed "real taste." At another level, gastronomy sought to reveal the essential truth about eaters. The most famous dictum of gastronomy indeed is "tell me what you eat, and I will tell you who you are."[21] Taste was supposed to reveal gender, class, national and civilizational identities, which were also understood in an essentialist and racialized manner. Whereas other discourses regulating food intake openly subordinated pleasure to something considered higher like health, morals or the soul, gastronomy subordinated pleasure to truth: the truth about foods and the truth of the self. According to Foucault, discourses are not mere linguistic representations but "practices that systematically build the objects of which they speak."[22] Thus, gastronomic discourse did not simply represent taste and much less uncovered its truth. Rather, gastronomy constructed the modern sense of taste as it has played an important role in shaping how modern societies approach the experience of taste. It also constructed essentialist and racialized subjectivities and hierarchies based on gustatory differences. The gastronomic construction of taste was contingent on the specific force field in the modern/colonial era. All discourses bring knowledge and power together,[23] and gastronomic discourse has served to articulate, create, reinforce and undermine power in the modern force field. Both the discourse of sexuality and the discourse of gastronomy are ways of policing bodies and pleasure through the establishment of rational expertise that decides what is normative and non-normative. Modern taste has been instrumental in the creation and policing of social and global hierarchies, as well as for their resistance.

The gastronomic focus on truth is intertwined not only with philosophical discourses but also with the power relations of capitalism, colonialism and imperialism. Gastronomic discourse established itself as the holder of truth regarding taste in the form of objective, scientific laws of cooking and eating. The pleasures secured from abiding by gastronomic laws had less to do with bodily and psychological affects than with the modern bourgeois and imperial subjectivities that they helped to constitute and perform. Gastronomy involved a discipline of consumption and a discipline of the sensing body. The bourgeois, and eventually all who aspired to a "civilized" and modern existence, submitted to this discipline in exchange for the non-bodily pleasure of asserting an identity that they saw as superior to others. Bourdieu has analyzed how the bourgeois culture of taste (in the broader meta-phorical aesthetic sense) marks social distinctions.[24] However, I argue that the role of gastronomy as a part of bourgeois culture goes beyond establishing and policing class and other distinctions inside any given society. Gastronomy has served to mark the distinctions that articulate the power structures of the modern/colonial global capitalist system.

The coloniality of modern taste refers to how taste as defined by gastronomy was shaped by the investment in turning the sense of taste into an instrument for the policing of global hierarchies of power by articulating racialized colonial dis-tinctions like Western/non-Western, modern/traditional, and civilized/primitive, which are the backbone of the global capitalist order. Aníbal Quijano coined the term "coloniality" to refer to the colonial character of contemporary structures of power that were established during the colonial era but endure after the formal demise of colonialism.[25] The bibliography of how colonial power structures shaped taste and cuisines in specific colonial and postcolonial contexts is growing.[26] What this book contributes to our understanding of the coloniality of taste is an explo-ration of how coloniality shaped gastronomic taste from the very beginning in Europe. Gastronomic notions of taste do not pre-exist the colonial order and were later imposed through it. It is my argument that the colonial order shaped how taste has been constructed in gastronomic thought from its inception. Coloniality informs the experience of taste in the centers of power, not just in the colonies.

Because of the modern/colonial power relations of global capitalism, the gastro-nomic notion of taste has affected and continues to affect most of the world in one form or another. This does not mean that all modern food cultures and cuisines are the same but that all modern societies have had to define their understanding of taste vis-a-vis gastronomic ideas that have been presented as a marker of cultural superiority like so many other modern capitalist ideas and institutions. Whenever a regional cuisine is packaged for global consumption as a "high cuisine," it under-goes a process of being adjusted to the taste preferences and culinary techniques enshrined by gastronomy.[27] French authors were the founders of the print expres-sion of gastronomic discourse, but they did so in the context of widespread trans-formations in Europe and in the world. These transformations most notably involve the rise of bourgeois culture and the development of global capitalism through Western imperial colonialism. This does not mean that gastronomy is universal, although it proposed itself as such. All it means is that it is the defining – but not the only – taste culture of the global capitalist order.

There are other modern discourses that regulate food intake. Nutrition, as a science and a morality that disciplines modern subjects through diet, has been rec-ognized as a mechanism of control.[28] Gastronomy, in comparison, is not imme-diately recognized as a disciplinary discourse because its stated focus is pleasure.

But, contrary to what the liberation hypothesis proposes, gastronomy controls taste pleasure quite tightly. It transformed a barebones notion of the sense of taste into an interface between bodies and the market. If nutrition is identified with state institutions like workhouses, prisons, schools and hospitals, gastronomy's main institutions are businesses like cooking schools, restaurants and food specialty shops. If nutrition produced docile and productive bodies, gastronomy produced voracious consumers of food products that keep the global capitalist system, and all its inequalities, functioning smoothly. Articulation with Western philosophy and the power relations of capitalism and imperialism is what made gastronomy more powerful than other taste cultures and what explains its global spread. It is also what makes it deeply problematic.

The Problem with Gastronomy

Gastronomic discourse established an affectless notion of taste as the foundation of a globally consequential sensory order that is oppressive of both those who subscribe to the gastronomic discipline and those who do not, although in significantly different ways. Through the processes of desensualization, bureaucratization and racialization discussed in this book, gastronomy turned the affectiveness of taste into an excess that had to be disavowed to achieve modern subjectivity. The affectless notion of taste enshrined by gastronomy as modern taste dramatically restricts the possibilities of sensory experience and bodily enjoyment of those who accept it because its prescriptive and standardized character is based on a dramatically limited understanding of the wide range of possibilities of the taste experience. Those embracing the affectiveness of taste, on the other hand, were racialized as inferior human beings. The personal, social and global colonial repression enacted by gastronomy is thus inscribed in the bodies of all modern subjects. In the case of those who have accepted gastronomic rule, sensory experience and physical pleasure are thwarted in the name of constituting a modern bourgeois "civilized" subjectivity and supposed racial superiority. In the case of those who do not know or do not accept gastronomic rule, their oppression only begins when they are classified as belonging to supposedly inferior "races" outside of the community of the modern and civilized. Gastronomic discourse, as an expression of modern culture, has played an active role in the construction and policing of enduring colonial and racial hierarchies between different peoples and different ways of cultivating the human sensorium. Believing that gastronomy liberated taste, or that it is some kind of natural human progression, keeps us beholden to an impoverished culture of taste that thwarts bodily enjoyment and continues to legitimize exploitative global power relations.

For taste to be rational according to the understanding of reason in Western thought in which mind and body are split and in which subjective and objective knowledge are distinct, many aspects of taste had to be repressed. The objective notion of taste advanced by gastronomy stands out when compared to the more fluid and affective notions of taste of other times and places. A comparison between the gastronomic and current ideas about taste in Western academic settings shows that the limitations of the gastronomic notion are finally becoming obvious in the West. In current scholarly conceptualizations of taste, what we normally call taste is considered as a multi-sensorial perception system in which taste, smell, touch, the trigeminal system, hearing and vision interact in multiple ways, and "flavor" is the term for the combination of all these systems unified by the act of eating.[29] Flavor is not a simple sensation but a perceptual modality, an active way of knowing.[30] Taste is a multi-sensorial experience that responds to biological, psychological and social variables. This concept of taste leaves behind basic tenets of modern thought, such as the separation and hierarchy of the senses, the distinction between body and mind, and the objectivity of taste. In current scholarly concepts of taste, the senses are interdependent. Taste is not separate from or inferior to the other senses but a player in a complex system. The senses do not belong to the domain of the body as separate from the mind because there is no distinction between mind and body. Thought is embodied in the brain and the neurological system. The brain is just another player in the perception and cognition system of taste. Finally, taste cannot possibly be objective because of the many biological, psychological and social variables involved in its experience. Taste as experience does not play out in exactly the same way every time, even in the same individual. While current scholarly notions of taste leave a lot of room for variations in the experience of taste, gastronomic discourse has been keen on downplaying and stigmatizing such variations in its quest to achieve the ideal of objectivity.

Cultures of taste informed by systems of thought that did not separate mind from body, and did not consider taste as a lower sense, had no need to construct taste as objective to combat its low status. Because of that, other cultures of taste have been more comfortable with the variability of the experience of taste and have given it a broader scope and flexibility than gastronomy. Gastronomy emerged in an effort to solve problems created by Western thought and, as Western thought is trying to solve these problems, the gastronomic notion of taste seems increasingly inadequate to Western scientists and thinkers.[31] The concept of taste in Western thought is now becoming more similar to the more complex and open-ended notions that have been the norm around the world. From this perspective, gastronomy can hardly be considered a pinnacle of progress. As a discourse intent on curtailing the richness of the experience of taste for the sake of achieving objectivity, gastronomy looks like a peculiar stumble in the global history of taste.

All cultures of taste have dealt with the idea that taste is an affective, embodied experience. One major peculiarity of the gastronomic notion of taste is its fear of affect. Affect is not the same as emotion. Affect is an intensity independent from language, whereas emotion is a qualified, narrativized intensity.[32] The pre- or para-linguistic character of affect does not reduce it to a "gut reaction," if we reject the Cartesian mind/body separation and accept that all the high capacities usually assigned to the mind are embodied.[33] The experience of affect is also not reducible to something individual and incommunicable, since experience has an important collective dimension. How our bodies affect and are affected by the world involves many variables, including socially and culturally specific ones. In the concept of taste as an affective experience, taste is the result of a combination of objective and subjective factors, it is not a given but the result of the collaboration of the food, the eater and the context. The food and the eater are both affecting and affected, and both are shaped by the interaction. Taste is a weave made of heterogeneous physiological, psychological, social and cultural strands that only come together in the act of eating, giving different results every time. In contrast, in the gastronomic notion, taste is predetermined in a way that is mostly unaffected by the eating subject. The eating subject is also conceived as unaffected by tasting. For gastronomy, the affective power of taste was an excess that had to be tamed if the notion of objective taste was to prevail.

Gastronomy is distinguished by its use of the written word for the specific purpose of disciplining the sense of taste. The use of the written word has been hailed as the key to the excellence of gastronomy.[34] However, this perspective is based on discredited theories regarding the superiority of literate cultures over oral ones. The written word could be employed in different ways, and it is neither a guarantee nor a hindrance for the elaboration of a sophisticated culture of taste. In the case of gastronomy the written word, particularly in print form, was the fundamental tool for articulating, disseminating and establishing gastronomic rule. Gastronomy's dependence on the printed word adds to its repressiveness. Language generally "dampens the intensity of affect,"[35] and the way in which gastronomy turned the experience of taste into something that should follow written rules reinforces this dampening. The modern gastronomic body is closed to affect, which is to say that it is closed to others and to the world. Contrasting with gastronomic notions, theorists of affect conceptualize the body as open. Bruno Latour, for example, sees the body not as a bounded substance but as an interface affected by many elements.[36] Food indeed has been characterized as an "alimentary agent" that affects us.[37] Limiting the experience of taste to capturing "the natural taste" and following "universal rules" negates the affective power of food and the richness inherent in the variability of the experience of taste. When following gastronomic rule, people are keener on knowing which foods they are supposed to like according to the experts than in exploring which ones would please them the most. They are also more concerned with performing a certain identity and elevated status than with having a transformative experience.

Gastronomic discourse as a nexus of knowledge and power in the modern/colonial era produced taste as we know it in contemporary modern societies. Lisa M. Heldke and Raymond D. Boisvert have argued that yesterday's philosophy is today's common sense.[38] Today even people unfamiliar with gastronomic texts routinely talk about taste in the ways proposed by gastronomic discourse. The apprehension about strong flavors overwhelming "the real taste of the food" and the dismissal of spicy flavors as unsuitable for refined palates is standard in contemporary food talk, echoing the ideas of the objectivity of taste and the supposed superiority of those who abstain from affective flavors. Embracing affective taste or any other deviation from the experience of taste that gastronomy made normative exposes people to be treated as less than modern and as lesser humans.

David Howes and Constance Classen have made the point that culture shapes not only how, and how much we see, but also what we see.[39] If culture also shapes how, how much and what we can taste, we can conclude that modern gastronomic culture allows us to taste precious little. Gastronomic subjects curbed their gustatory pleasure according to rules but found new pleasures in ascertaining a supposedly superior social and racial identity when establishing and following those rules. However, gastronomy's downside goes beyond the loss inherent in modern self-discipline and internalized repression. Howes and Classen also argue that our ways of sensing affect both how we engage with the environment and how we engage with each other.[40] Gastronomy has structured not only modern ways of tasting but also modern ways of relating to each other, which include class, gender, ethnic, racial, cultural and other modern/colonial hierarchies. The need to establish and police colonial differences between peoples constructed as traditional and modern, Oriental and Western, primitive and civilized had a constitutive and enduring impact on the modern gastronomic notion of taste. Our sensory enjoyment is the poorer for it.

It might be argued that gastronomy has overcome coloniality, given that its horizons since the late twentieth century have been extended to incorporate ingredients, dishes and techniques from all over the world. However, the coloniality of the unequal relationship between different peoples and different ways of experiencing taste cannot be leveled by only inclusion. The way in which contemporary "foodies" relate to other peoples and cultures is not very different from what was established in the nineteenth century. A study by Josée Johnston and Shyon Baumann found that the multicultural foodie discourse reinforces class distinctions, in spite of its egalitarian pretensions. They concluded that in foodie discourse, knowledge of the food of the other is a way of asserting middle-class membership.[41] Similarly, Lisa Heldke has argued that "food adventurism" has philosophical colonialist underpinnings because it exploits the other as a resource in an unequal power relationship.[42] These and other studies demonstrate that overcoming the coloniality of modern taste cannot be achieved by inclusion. Inclusion continues the unequal colonial power relationship between those with the power to include and those who can only hope to be included in a subordinated way, and who are included only according to rules established without their participation.[43]

The expansion of the set of ingredients, dishes, recipes and techniques from different parts of the world accepted by gastronomy today serves the capitalist market's need for constant innovation, but it does not change the unequal power relations that make Europeans and Euro-Americans the protagonist subjects who decide what elements of other culinary knowledges are valuable or can be considered "universal" or "global." The insatiable hunger for novelty is indeed one of the most damaging characteristics of modern taste. The gastronomic and foodie taste for innovation fuels unsustainable and exploitative food production practices. The resources, labor and knowledge of the vast regions of the world that are not beneficiaries of the modern/colonial global capitalist order continue to be exploited for the enjoyment of those that modernity and gastronomy have made believe that they are entitled to and deserving of the alimentary wealth of the planet.

Many scholars have been attentive to the need of recognizing that the peoples and places marginalized by modernity have nevertheless had a significant impact on the development of global modern taste, in many ways resisting and subverting the normativity of the gastronomic notion of taste. The work of Sidney Mintz has been exemplary on this account. Beyond showing how Caribbean slave-produced sugar fueled the industrial revolution,[44] he exposed the need to explore and acknowledge in full the contributions of Asia to world cuisine.[45] Mintz also made the point that enslaved Africans shaped the taste and culinary cultures of the greater Caribbean region.[46] More recently, Krishnendu Ray has analyzed how "ethnic restaurateurs" engage and negotiate with the rules and rankings established by the dominant taste culture in the United States. He shows the agency that "ethnic restaurateurs" have as tastemakers, even when they continue to be marginalized materially and discursively.[47] Many other scholars have focused on the contributions of specific peoples and cultures to the current global cultures of taste. The existence of studies like these suggests that there is an awareness of the colonial and Eurocentric character of gastronomy and gastronomic scholarship, and a desire to move beyond it. However, in order to overcome the coloniality of gastronomy and its scholarship more fully, we need to go beyond the goals of inclusion and recognition. The coloniality of gastronomy cannot be overcome without balancing the power ratio between different peoples, knowledges and sensoria. A small but important step toward the overcoming of the limitations of the modern/colonial approach to taste is to allow perspectives coming from different taste cultures and histories to disrupt and invalidate the megalomaniac narratives of modernity and gastronomy. I offer this book as a step in that direction.

This introduction has laid out the general argument of the book, which is expanded and illustrated in the rest of the book. Chapter 1: "The Narrative of Gastronomic Progress" develops a critique of the historical narrative that structures much of the scholarship on gastronomy and how it aggrandizes the global significance of gastronomy through the misrepresentation and devaluation of other cultures of taste. The chapter discusses how this narrative adopts a Eurocentric perspective that universalizes the contingencies of the European experience, how its conflation of space and time transforms the taste culture of other places into the taste culture of past historical periods, and how its representation of the West as protagonist of the global history of taste is reinforced by the portrayal of other peoples as prisoners of their local cultures. The chapter argues that the narrative of gastronomy does not just innocently celebrate a culinary culture that is dear to many but also continues to naturalize and reinforce modern/colonial inequalities.

The next three chapters are devoted to the three main processes that I argue have shaped the modern gastronomic notion of taste: desensualization, bureaucratization and racialization. These processes are analyzed as the product of specific contingencies of the modern/colonial European context. The argument is based on an analysis of foundational texts of gastronomy, including classic gastronomic books, cookbooks, periodicals, magazines and almanacs. These texts were written mostly in France and Britain, but many of them circulated in other

European countries and in the United States. Rather than focusing on the biography and works of any specific writer, I present the modern gastronomic notion of taste as the collective construction of many different authors.

Chapter 2: "Desensualizing Taste" places gastronomic texts in the context of the philosophical debates to which they responded and analyzes gastronomic texts as a site of struggle. The analysis shows the ambivalence of the gastronomic genre, how it reached a compromise between its interest in validating food and taste as a legitimate field of knowledge, on the one hand, and the limitations that rationalism and aesthetics placed on taste and sensing, on the other. The chapter argues that this compromise led gastronomy to strip its notion of taste of the sensuous subjective and affective aspects that exceed the limits of rationalist objectivity. It concludes with a discussion of how gastronomic texts used self-deprecating humor to express their resentment against a view of knowledge that excludes pleasurable sensing.

Chapter 3: "Bureaucratizing Taste" shows how gastronomic writers employed the printed word to establish themselves as bureaucrats of taste who ushered in the transition from aristocratic taste culture to a culture of taste defined by the modern/colonial capitalist market. It shows how they fashioned the consumption of the latest food commodities approved by them as participation in the march of modern progress. The chapter argues that capitalism gave the sense of taste relative freedom from moral and aesthetic injunctions against excess, but it also led to a process of standardization of cooking and eating to make the process of consumption more efficient. It concludes that gastronomy used print to standardize the taste experience with rules that restrict the exploration of the wide range of possibilities of the experience of taste.

Chapter 4: "Racializing Taste" shows how the shift away from spices in European cuisines was the result of a long and active engagement of gastronomy with colonial race discourse. It argues that the gastronomic rejection of spices represents the will to control the affective power of taste and that it helped to construct the idea of a distinct and supposedly superior white European race. The chapter discusses how gastronomic writing changed the meaning of spices from the almost unattainable luxury of faraway high civilizations to the alluring but dangerous sense of taste of peoples that modernity marginalized. The chapter concludes that gastronomic discourse added a sensorial dimension to modern racist thought by turning openly sensual and affective taste into the taste of so-called inferior races. The gastronomic enjoyment of affective foods and substances was possible, but it was framed as an exciting and dangerous transgression. Gastronomic writers established a racialized understanding of culinary and gustatory diversity as if it were objective scientific knowledge.

The next chapter rounds off the analysis of the coloniality of modern taste by looking beyond it. Chapter 5: "Taste, Otherwise" demonstrates the relative poverty of the gastronomic notion of taste by presenting an overview of how taste was conceptualized in classic Arabic, Chinese, Indic, Nahua and Yorùbá thought. These systems of thought differ in their approaches to taste, but none of them restricted the experience of taste as much as modern gastronomy. The chapter demonstrates how the notions of epistemology, aesthetics and ethics found in these five systems of thought enabled conceptualizations of taste that embraced to different degrees its epistemic capabilities without discounting its subjectivity and affective power. The chapter concludes that, when compared to notions of taste enabled by systems of thought other than the modern one, the gastronomic notion of taste comes across as peculiarly bereft of affect.

The conclusion of the book, "The Gustatory Logic of Consumer Capitalism," shows how the approach to taste that originated in nineteenth-century gastronomic thought has continued to develop and adapt to the latest manifestations of capitalism and coloniality. This chapter discusses how the processes of desensualizing and bureaucratizing taste have intensified through the use of audiovisual and food and food flavor technologies and argues that the racialization of taste continues to fuel the relentless capitalist commodification that exploits nature, cultures and peoples. The conclusion ends with a discussion of the kinds of movements and practices that reflect a yearning for more affective approaches to taste that could open paths toward the decolonization of the gustatory experience.

Notes

1 David Howes, ed., *Empire of the Senses: The Sensual Culture Reader* (Oxford; New York: Berg, 2005), 3.
2 Walter Mignolo has developed this idea extensively. See, for example, Walter D. Mignolo, "Delinking: The Rhetoric of Modernity, the Logic of Coloniality and the Grammar of De-Coloniality," *Cultural Studies* 21, no. 2–3 (March 2007): 449–514, https://doi.org/10.1080/09502380601162647.
3 Enrique D. Dussel, "Beyond Eurocentrism: The World-System and the Limits of Modernity," in *The Cultures of Globalization*, eds. Fredric Jameson and Masao Miyoshi (Durham, NC: Duke University Press, 1999), 3–31.
4 Walter D. Mignolo, *The Darker Side of Western Modernity: Global Futures, Decolonial Options* (Durham, NC: Duke University Press, 2011), 66.
5 This exemplifies what Santiago Castro-Gómez has called the hubris of the zero point. The zero point is the location of the sovereign perspective adopted by Enlightenment thinkers to look at the world without being seen. This allowed them to interpret the world according to their imperial interests while being convinced of their neutrality. *La hybris del punto cero: Ciencia, raza e ilustración en la Nueva Granada (1750–1816)* (Bogotá: Editorial Pontificia Universidad Javeriana, 2005), 18.
6 Jean Louis Flandrin and Massimo Montanari, *Food: A Culinary History from Antiquity to the Present*, ed. Albert Sonnenfeld (New York: Penguin Putnam, 2000), 418–432.
7 Priscilla Parkhurst Ferguson, *Accounting for Taste: The Triumph of French Cuisine* (Chicago: University of Chicago Press, 2004), 24.
8 For a discussion of the relative autonomy of the cultural field see Pierre Bourdieu, *Distinction: A Social Critique of the Judgement of Taste*, trans. Richard Nice (Cambridge, MA: Harvard University Press, 1984).
9 Michel Foucault, *The History of Sexuality: An Introduction* (New York: Vintage Books, 1990).
10 Michel Foucault, *The Foucault Reader*, ed. Paul Rabinow (New York: Pantheon Books, 1984), 356–357.
11 G.J.H. van Gelder, *God's Banquet: Food in Classical Arabic Literature* (New York: Columbia University Press, 2000), 23.
12 Charles Malamoud and David White, *Cooking the World: Ritual and Thought in Ancient India* (Oxford: Oxford University Press, 1998).

13 Carolyn Korsmeyer, *Making Sense of Taste: Food & Philosophy* (Ithaca, NY: Cornell University Press, 1999), 11–37.

14 Viktoria von Hoffmann, *From Gluttony to Enlightenment: The World of Taste in Early Modern Europe* (Urbana, IL: University of Illinois Press, 2016), loc. 1555 of 7807, Kindle.

15 Hoffmann, loc. 952 of 7807, Kindle.

16 Hoffmann, loc. 2504–2509 of 7807, Kindle.

17 E.C. Spary, *Eating the Enlightenment: Food and the Sciences in Paris, 1670–1760*, eds. Maxine Berg and Helen Clifford (Chicago: University of Chicago Press, 2014), loc. 4441 of 9350, Kindle.

18 Jean Anthelme Brillat-Savarin, *The Physiology of Taste, or, Meditations on Transcendental Gastronomy*, trans. M.F.K. Fisher (New York: Vintage Books, 2011).

19 E.C. Spary, "Making a Science of Taste: The Revolution, the Learned Life, and the Invention of 'Gastronomie'," in *Consumers and Luxury: Consumer Culture in Europe 1650–1850* (Manchester: Manchester University Press, 1999), 180.

20 Foucault, *History of Sexuality*, 56.

21 Brillat-Savarin, 15.

22 Michel Foucault, *The Archaeology of Knowledge; and the Discourse on Language* (New York: Pantheon Books, 1982), 49.

23 Foucault, *History of Sexuality*, 100–101.

24 Bourdieu.

25 Aníbal Quijano, "Coloniality and Modernity/Rationality," *Cultural Studies* 21, no. 2 (March 2007): 168–178, https://doi.org/10.1080/09502380601164353.

26 See for example Adolfo Albán Achinte, *Sabor, poder y saber: Comida y tiempo en los valles afroandinos del Patía y Chota-Mira* (Popayán, Colombia: Editorial Universidad del Cauca, 2015); María Elena García, *Gastropolitics and the Specter of Race: Stories of Capital, Culture, and Coloniality in Peru* (Oakland, CA: University of California Press, 2021); Parama Roy, *Alimentary Tracts: Appetites, Aversions, and the Postcolonial* (Durham, NC: Duke University Press, 2010); Marcy Norton, *Sacred Gifts, Profane, Pleasures: A History of Tobacco and Chocolate in the Atlantic World* (Ithaca, NY: Cornell University Press, 2008); Rebecca Earle, *The Body of the Conquistador: Food, Race, and the Colonial Experience in Spanish America, 1492–1700* (Cambridge, UK; New York: Cambridge University Press, 2013); and E.M. Collingham, *Imperial Bodies: The Physical Experience of the Raj, c. 1800–1947* (Cambridge, UK; Malden, MA: Polity Press; Blackwell Publishers, 2001), among others.

27 I develop this point in Zilkia Janer, "(In)Edible Nature: New World Food and Coloniality," *Cultural Studies* 21, no. 2 (March 2007): 385–405, https://doi.org/10.1080/09502380601162597.

28 John Coveney, *Food, Morals and Meaning* (London: Routledge, 2006).

29 Malika Auvray and Charles Spence, "The Multisensory Perception of Flavor," *Consciousness and Cognition* 17, no. 3 (September 2008): 1026, https://doi.org/10.1016/j.concog.2007.06.005.

30 Auvray and Spence, 1025.

31 See for example Raymond D. Boisvert and Lisa M. Heldke, *Philosophers at Table: On Food and Being Human* (London: Reaktion Books, 2016).

32 Brian Massumi, "The Autonomy of Affect," *Cultural Critique* Autumn, no. 31 (1995): 88, https://doi.org/10.2307/1354446.

33 Massumi, 90.

34 Ferguson, 92.

35 Massumi, 86.

36 Melissa Gregg and Gregory J. Seigworth, eds., *The Affect Theory Reader* (Durham, NC: Duke University Press, 2010), 11.

37 Ben Highmore, "Alimentary Agents: Food, Cultural Theory and Multiculturalism," *Journal of Intercultural Studies* 29, no. 4 (November 2008): 381–398, https://doi.org/10.1080/07256860802372337.

38 Boisvert and Heldke, loc. 2121 of 3400, Kindle.

39 David Howes and Constance Classen, *Ways of Sensing: Understanding the Senses in Society* (New York: Routledge, 2014), 1.
40 Howes and Classen, 6.
41 Josée Johnston and Shyon Baumann, *Foodies: Democracy and Distinction in the Gourmet Foodscape* (London: Routledge, 2009).
42 Lisa Heldke, *Exotic Appetites: Ruminations of a Food Adventurer* (London: Routledge, 2003).
43 Walter D. Mignolo, *The Darker Side of Western Modernity: Global Futures, Decolonial Options* (Durham, NC: Duke University Press, 2011), xv.
44 Sidney W. Mintz, *Sweetness and Power: The Place of Sugar in Modern History* (New York: Penguin Books, 1986).
45 Sidney Mintz, "Asia's Contributions to World Cuisine: A Beginning Inquiry," in *Food and Foodways in Asia: Resource, Tradition and Cooking,* ed. Sidney C.H. Cheung and Chee Beng Tan (London: Routledge, 2007), 201–210.
46 Sidney Wilfred Mintz, *Tasting Food, Tasting Freedom: Excursions into Eating, Culture, and the Past* (Boston: Beacon Press, 1996), 33–49.
47 Krishnendu Ray, *The Ethnic Restaurateur* (London: Bloomsbury Academic, 2016).

References

Albán Achinte, Adolfo. *Sabor, poder y saber: Comida y tiempo en los valles afroandinos del Patía y Chota-Mira.* Popayán, Colombia: Editorial Universidad del Cauca, 2015.
Auvray, Malika, and Charles Spence. "The Multisensory Perception of Flavor." *Consciousness and Cognition* 17, no. 3 (September 2008): 1016–1031. https://doi.org/10.1016/j.concog.2007.06.005.
Boisvert, Raymond D., and Lisa M. Heldke. *Philosophers at Table: On Food and Being Human.* London: Reaktion Books, 2016.
Bourdieu, Pierre. *Distinction: A Social Critique of the Judgement of Taste.* Translated by Richard Nice. Cambridge, MA: Harvard University Press, 1984.
Brillat-Savarin, Jean Anthelme. *The Physiology of Taste, or, Meditations on Transcendental Gastronomy.* Translated by M.F.K. Fisher. New York: Vintage Books, 2011.
Castro-Gómez, Santiago. *La hybris del punto cero: Ciencia, raza e ilustración en la Nueva Granada (1750–1816).* Bogotá: Editorial Pontificia Universidad Javeriana, 2005.
Collingham, E.M. *Imperial Bodies: The Physical Experience of the Raj, c. 1800–1947.* Cambridge, UK; Malden, MA: Polity Press; Blackwell Publishers, 2001.
Coveney, John. *Food, Morals and Meaning.* London: Routledge, 2006.
Dussel, Enrique D. "Beyond Eurocentrism: The World-System and the Limits of Modernity." In *The Cultures of Globalization.* Edited by Fredric Jameson and Masao Miyoshi, 3–31. Durham, NC: Duke University Press, 1999.
Earle, Rebecca. *The Body of the Conquistador: Food, Race, and the Colonial Experience in Spanish America, 1492–1700.* Cambridge, UK; New York: Cambridge University Press, 2013.
Ferguson, Priscilla Parkhurst. *Accounting for Taste: The Triumph of French Cuisine.* Chicago: University of Chicago Press, 2004.
Flandrin, Jean Louis, and Massimo Montanari. *Food: A Culinary History from Antiquity to the Present.* Edited by Albert Sonnenfeld. New York: Penguin Putnam, 2000.
Foucault, Michel. *The Archaeology of Knowledge; and the Discourse on Language.* New York: Pantheon Books, 1982.
———. *The Foucault Reader.* Edited by Paul Rabinow. New York: Pantheon Books, 1984.
———. *The History of Sexuality: An Introduction.* New York: Vintage Books, 1990.
García, María Elena. *Gastropolitics and the Specter of Race: Stories of Capital, Culture, and Coloniality in Peru.* Oakland, CA: University of California Press, 2021.

Gelder, G.J.H. van. *God's Banquet: Food in Classical Arabic Literature*. New York: Columbia University Press, 2000.

Gregg, Melissa, and Gregory J. Seigworth, eds. *The Affect Theory Reader*. Durham, NC: Duke University Press, 2010.

Heldke, Lisa. *Exotic Appetites: Ruminations of a Food Adventurer*. London: Routledge, 2003.

Highmore, Ben. "Alimentary Agents: Food, Cultural Theory and Multiculturalism." *Journal of Intercultural Studies* 29, no. 4 (November 2008): 381–398. https://doi.org/10.1080/07256860802372337.

Hoffmann, Viktoria von. *From Gluttony to Enlightenment: The World of Taste in Early Modern Europe*. Urbana, IL: University of Illinois Press, 2016.

Howes, David, and Constance Classen. *Ways of Sensing: Understanding the Senses in Society*. London: Routledge, 2014.

Janer, Zilkia. "(In)Edible Nature: New World Food and Coloniality." *Cultural Studies* 21, no. 2 (March 2007): 385–405. https://doi.org/10.1080/09502380601162597.

Johnston, Josée, and Shyon Baumann. *Foodies: Democracy and Distinction in the Gourmet Foodscape*. London: Routledge, 2009.

Korsmeyer, Carolyn. *Making Sense of Taste: Food & Philosophy*. Ithaca, NY: Cornell University Press, 1999.

Malamoud, Charles, and David White. *Cooking the World: Ritual and Thought in Ancient India*. Oxford: Oxford University Press, 1998.

Massumi, Brian. "The Autonomy of Affect." *Cultural Critique* Autumn, no. 31 (1995): 83. https://doi.org/10.2307/1354446.

Mignolo, Walter D. "Delinking: The Rhetoric of Modernity, the Logic of Coloniality and the Grammar of de-Coloniality." *Cultural Studies* 21, no. 2–3 (March 2007): 449–514. https://doi.org/10.1080/09502380601162647.

———. *The Darker Side of Western Modernity: Global Futures, Decolonial Options*. Durham, NC: Duke University Press, 2011.

Mintz, Sidney. "Asia's Contributions to World Cuisine: A Beginning Inquiry." In *Food and Foodways in Asia: Resource, Tradition and Cooking*. Edited by Sidney C.H. Cheung and Chee Beng Tan, 201–210. London: Routledge, 2007.

Mintz, Sidney W. *Sweetness and Power: The Place of Sugar in Modern History*. New York: Penguin Books, 1986.

Mintz, Sidney Wilfred. *Tasting Food, Tasting Freedom: Excursions into Eating, Culture, and the Past*. Boston: Beacon Press, 1996.

Norton, Marcy. *Sacred Gifts, Profane, Pleasures: A History of Tobacco and Chocolate in the Atlantic World*. Ithaca, NY: Cornell University Press, 2008.

Quijano, Aníbal. "Coloniality and Modernity/Rationality." *Cultural Studies* 21, no. 2 (March 2007): 168–178. https://doi.org/10.1080/09502380601164353.

Ray, Krishnendu. *The Ethnic Restaurateur*. London: Bloomsbury Academic, 2016.

Roy, Parama. *Alimentary Tracts: Appetites, Aversions, and the Postcolonial*. Durham, NC: Duke University Press, 2010.

Spary, E.C. *Eating the Enlightenment: Food and the Sciences in Paris, 1670–1760*. Chicago: University of Chicago Press, 2014.

———. "Making a Science of Taste: The Revolution, the Learned Life, and the Invention of 'Gastronomie'." In *Consumers and Luxury: Consumer Culture in Europe 1650–1850*. Edited by Maxine Berg and Helen Clifford, 170–182. Manchester: Manchester University Press, 1999.

1
THE NARRATIVE OF GASTRONOMIC PROGRESS

The core of gastronomic discourse is a historical narrative of progress starring France as culinary leader of the West.[1] The narrative of gastronomic progress was established in classic gastronomic writings, and it continues to be the backbone of most academic and popular accounts of the history of food and taste. In the process of defining gastronomy, this narrative also defined all other cultures of taste as lacking the superior characteristics that gastronomy is supposed to have. A historical narrative is a story that connects and gives meaning to events. The meaning of historical events is not inherent in them; it is constituted in the structure and language of the narrative.[2] The foundational narrative of gastronomy cast gastronomy as the first culture of taste to acquire certain characteristics that made all other ones outdated. This is not just a harmless Eurocentric ego-trip, it is an example of the coloniality of knowledge. The narrative of gastronomic progress is an expression of the European paradigm of rational knowledge which, in the words of Aníbal Quijano, "was not only elaborated in the context of, but as a part of, a power structure that involved the European colonial domination over the rest of the world."[3]

The diffusion power of print and the imperial power of Europe turned the narrative of gastronomic progress into a template used to evaluate all past and present cultures of taste. The great variety of cultures of taste across time and space has been inserted into the master narrative of gastronomy, rather than being understood on their own terms. The narrative of gastronomy continues to inform all kinds of food and taste-related scholarship. As Dipesh Chakrabarty put it, in academic historical discourse "'Europe' remains the sovereign, theoretical subject of all histories, including the ones we call 'Indian,' 'Chinese,' 'Kenyan' and so on."[4] Accounts of the local, regional or global history of food, taste and cuisines are conceived using the narrative of gastronomy as a standard. Cultures of taste other than gastronomy are portrayed as stagnant and lacking if they do not have the modern characteristics

DOI: 10.4324/9781003331834-2

established by gastronomy as if they were a desirable universal destiny. The narrative of gastronomy created the myth of the superiority of modern taste, as defined in gastronomic writing, through the misrepresentation and marginalization of peoples with different ways of conceptualizing and experiencing taste.

The violence of the way in which the narrative of gastronomic progress established gastronomy as the first culture of taste worthy of the name becomes obvious with even a quick survey of the great diversity of sophisticated cultures of taste in every region of the world before the emergence of gastronomy. The written record available for the study of taste cultures before the modern period is spotty at best. However, the available records allow us to glimpse into what undoubtedly were full-fledged cultures of taste. In the Middle East, people developed a sophisticated culture of taste as early as the second millennium BCE. The Yale Culinary Tablets contain a collection of recipes from southern Babylonia c.1700 BCE. Experts consider that these tablets were a part of a much larger collection, perhaps even a large library, dedicated to cooking-related knowledge.[5] The recipes offer a window into a refined way of cooking that employed different techniques to enrich the flavor, texture, appearance and aroma of food. In the Middle Ages, cuisine reached high levels of sophistication in the Islamic world. A high point of this sophistication was reached in ninth-century Baghdad, where the fashionable were obsessed not only with fancy eating but also with reading and writing about the culinary arts.[6] Throughout the ninth and tenth centuries, cookbooks, dining etiquette manuals, and food- and dining-related poetry were widely popular in Baghdad. Ziryab, a freed slave and musician from Baghdad, moved to Córdoba and spread the dining habits of Baghdad, ultimately becoming an arbiter of taste. He set standards for etiquette, fashion and dining, and his influence on Spanish and European dining habits can still be seen today.[7] Indeed, the influence of the sophisticated culture of taste of Medieval Islam can be seen today all over the world.

China also developed a sophisticated taste culture, millennia before the emergence of French gastronomy. Imperial Chinese taste culture is distinguished by the knowledge and incorporation of an extremely wide variety of foods, and by an abundant food-related literature. Theorizing about food and taste in China is documented as early as the second millennium BCE, when Yi Yin (1648–1549 BCE) laid out a culinary theory and a classification of foods of lasting influence.[8] In the Sung-Yüan period (960–1368), flourishing trade allowed the elite and middle class to develop an incredibly varied and technically sophisticated cuisine.[9] During the late Ming period, consumer culture boomed and the writing of culinary authors like Xu Wei, Zhang Dai and Gao Lian shows the importance given to the refined pleasures of food.[10] Culinary literature continued to grow throughout the seventeenth and eighteenth centuries. One of the best know texts is *Recipes from Sui Garden*, published by Yuan Mei in 1796.[11] Chinese culinary authors described and prescribed in matters related to cooking, eating, etiquette and the good life in general.

In South India, food was already rich and varied in the first few centuries CE.[12] Rice was processed and prepared in many different ways – including the *āppam*, *idi-āppam*, *dōsai* and *adai* – that can still be recognized today as the region's signature dishes.[13] Around the start of the common era, writers Charaka, Sushrutha and Vāghbhata codified Hindu medical ideas.[14] These texts include not only theories about hot and cold foods and principles of Ayurvedic medicine but also theories of taste. Charaka, for example, noted six pure tastes and 63 mixed tastes,[15] and the *Sushrutha Samhitā* is largely a cookery book.[16] Many literary texts contain descriptions of meals that allow us to appreciate the importance given to well-appointed food in different parts of India in different time periods. The ancient Sanskrit texts *Ramayana* and *Mahabharata* mention a variety of rich dishes of meat, rice and spices.[17] In South India, there is an exhaustive book on cooking, called the *Supa Shastra*, from 1516 CE.[18] Writings on food in Kannada language go back to at least 920 CE, which document the persistence and gradual evolution of food items in this specific region of India.[19] Sixteenth-century Mughlai cuisine, mostly in contemporary India and Pakistan, is documented in memoirs like the *Ain-i-Akbari*, which contains descriptions of the great assortment of foods and dishes prepared for the court by cooks from different countries.[20] This memoir also includes a theory of flavors and an account of plants used for perfumes.[21] The coming together and further elaboration of the great cuisines of the Islamic world with the local foods, cuisines and culinary knowledge of India produced the complex Mughlai cuisine, which is only the best known of the many cuisines of South Asia.

Indigenous civilizations in the Americas, like the Aztec, Inca and Maya, developed elaborate and distinctive taste cultures making use of the abundant and varied food resources from vast regions. The food prepared by the Aztecs for festivals, the royal household and the markets was described by the Spanish Friar Bernardino de Sahagún in the sixteenth century.[22] There were many different kinds of tamales and stews, each one with a different combination of vegetables, meats and chilies. The attention to quality can be seen in the description of the finer tortillas and meals prepared for the court[23] and in the spelling out of criteria to distinguish the good tamal, tortilla and bread vendors from the bad ones.[24] Corn was prepared in many different ways all over the American continent, in many cases involving sophisticated techniques like nixtamalization, which implies soaking corn in slaked lime before grinding. This technique produces a pliable dough that is more nutritious than untreated corn and has many more culinary applications. The importance of corn as a culinary staple and canvas for creativity was extended to artistic representations and the religious realm. There was a clearly defined theory of taste and cooking.

There are enough indications that the peoples of Africa gave considerable importance to the cultivation of taste for centuries. Medieval Arabic texts contain descriptions of the foods and foodways of West African peoples in the Middle Ages. A compilation of these sources reveals that West Africans developed food processing and cooking techniques for a wide variety of foodstuffs.[25] European travelers from as early as the seventeenth century were amazed by excellent food preparations and lavish feasts.[26] These travelers give only a glimpse of what no doubt were complete and sophisticated culinary and taste cultures.

Taste culture in medieval Europe, long vilified from the perspective of modern gastronomy as unrefined and ostentatious, has been the focus of increased interest.

Culinary historians in the past few decades have demonstrated that this cuisine was complex, subtle and varied, and had a particular aesthetic sensibility.[27] A similar reevaluation needs to be undertaken regarding all the taste cultures that have been marginalized by the narrative of gastronomic progress. There is ample evidence of the existence of sophisticated cultures of taste all over the world throughout centuries. The above mentioned and many other cultures of taste that predate the emergence of gastronomy contain all of the features that supposedly distinguish gastronomy from previous taste cultures. Features like refinement, innovation, codification, cosmopolitanism, self-consciousness, intellectualization and aestheticization are by no means an original contribution of gastronomy to the global history of taste. The technical and aesthetic sophistication of the taste cultures that preceded gastronomy by centuries should be hard to dismiss, even though the historical records are in most cases incomplete. However, the narrative of gastronomic progress marginalized them all by pointing out the ways in which they were different from gastronomy, and transforming those differences into shortcomings: they were not written or printed, they were related to religion or medicine, they were for elites only, they were not national, they were not class inflected, and so on. Taking the contingent qualities of gastronomy as if they were a universal point of arrival has reduced other cultures of taste to interesting but minor episodes in an evolutionary history that led ineluctably to gastronomy. Gastronomy does have distinctive characteristics, as all taste cultures do, but these characteristics are contingent to a specific geopolitical context and not the necessary destiny of human taste cultures. To understand the specificity of gastronomy without presupposing its superiority, we need to actively reject the epistemic Eurocentrism of modernity.

Eurocentrism, unlike most other ethnocentrisms, has had the power to impose its views on the outsiders that it subordinates. As Walter Mignolo explains, modernity/coloniality created Europe as the center of space and the point of arrival in time, and it has also given it the epistemic privilege of being the locus of enunciation, a privilege that it retains even today that Europe is no longer the center of world capitalism.[28] The epistemic privilege of Europe continues to orient the understanding of gastronomy and other cultures of taste. Many scholars acknowledge the limitations of using the narrative of gastronomy as a template to understand the history of other cultures of taste but continue to do so nonetheless. Even in the case of the taste culture of the powerful United States, breaking with the Eurocentric perspective has been an elusive task.

In a report commissioned by the James Beard Foundation, an organization whose stated mission used to be "to celebrate, nurture and honor America's diverse culinary heritage,"[29] the authors timidly challenge the presumed universality of the standards defined by the French:

> it seems reductive to confine the richness and diversity of influences that are found in the foods we eat into one cuisine—unless, of course, we distance ourselves enough from the French idea of cuisine to create a language unique to the United States.[30]

They see the inadequacy of the concepts defined by the French experience, yet the paper is an attempt to apply the French concept of cuisine to the United States, rather than an effort to develop a specific language for the United States as they rightly suggest is needed. It is crucial to understand that a different set of concepts is needed to talk about each specific culinary culture and not only about the culinary culture of the United States. The point is not that the United States is exceptional and therefore it needs its own language but rather that all culinary cultures contain their own concepts and theories, which are obliterated when the universal applicability and desirability of French concepts are accepted. This leads to the misunderstanding and subalternization of culinary and taste cultures that are not assimilable into the French pattern. The diffidence of the James Beard report is understandable. Challenging the epistemic privilege of Europe is a difficult task, since it involves upending so many modern beliefs. However, it is a necessary challenge if we aspire to the decolonization of knowledge and taste.

Scholars who have tried to present more comprehensive global histories of food and taste have faced difficulties because the historical record of taste cultures around the world does not fit nicely into the established narrative of gastronomic progress. Jeffrey Pilcher, for example, starts his book *Food in World History* with the premise that modernization is characterized by a movement toward simpler flavors. However, he introduces a crucial qualification: "Europeans were not the only ones to follow this path, nor did it represent an exclusive route to culinary modernity."[31] After discussing cuisine and modernization in different parts of the world he concludes:

> Theorizing connections between nouvelle cuisines and social modernization becomes even more difficult as the examples multiply. The few available cooking manuscripts from the eighteenth-century Middle East indicate that foods may have been undergoing a comparable process of simplification, but at a time of extremely limited social change. Mexico, although influenced by the political and economic transformations of the Enlightenment, did not abandon the spicy, complex chile pepper stews of the colonial period in favor of simpler dishes fashionable in Europe. Thus, the associations discussed in this chapter must necessarily remain speculative until more data become available from around the world.[32]

Considering more examples of taste cultures makes theorizing more difficult only if we refuse to give up the narrative of modernity as the key to understanding world history. The available data already make it clear that there is nothing new, exceptional or superior about the preference for simpler flavors. Different peoples have espoused or eschewed this preference for different reasons at different times. While new information would be welcome, what we need is not new information as much as new decolonial ways of understanding the information that we have. A non-Eurocentric approach to the existing knowledge should lead to the delegitimization of the standard modern historical narrative of gastronomy.

The use of gastronomy as the model to understand all cultures of taste is an example of one of the main strategies of the modern/colonial control of knowledge and subjectivity, which Santiago Castro-Gómez has called a "zero-point epistemology." According to Castro-Gómez, Europe gained epistemological hegemony over all the other cultures of the world by identifying European particularity with universality. In so doing,

> The coexistence of diverse ways of producing and transmitting knowledge is eliminated because now all forms of human knowledge are ordered on an epistemological scale from the traditional to the modern, from barbarism to civilization, from the community to the individual, from the orient to the occident.[33]

Zero-point epistemology does not allow us to fully engage with the rich diversity of cultures of taste across time and space. Gastronomic scholarship is still molded by modernity's understanding of itself, even while recognizing its limits. This is not due to the shortcomings of any individual scholar. Rather it is proof that epistemic Eurocentrism continues to serve as unexamined common sense. We need to make a self-conscious effort to understand the diversity of cultures of taste without taking the presumed superiority, inevitability and desirability of Western modernity and gastronomy as a starting point. That presupposition is the founding stone of the narrative of gastronomic progress and it is what allowed for gastronomy to be undeservedly established as the foremost culture of taste.

The narrative of gastronomy also follows a style of representation that Fernando Coronil called "occidentalist," which produces polarized and hierarchical conceptions of the West and its Others. Occidentalist conceptions separate the components of the world into bounded units, disaggregate their relational histories, turn differences into hierarchies, naturalize these representations and thus intervene in the reproduction of unequal power relations.[34] An analysis of the formal narrative elements of the narrative of gastronomy allows us to see how the historiography of taste cultures has created a larger-than-life representation of modern taste culture by othering and marginalizing others. The major narrative elements concern the adoption of a Eurocentric voice and perspective, the conflation of space and time, and the characterization of the West as the main agent in the global history of taste.

Eurocentric Voice and Perspective

The voice of the narrative of gastronomy takes on an objective and authoritative tone as if it were dispensing universal truths, but it presents gastronomy and other taste cultures from the provincial perspective of modern European men, who wanted to see themselves as members of a superior civilization. One simple but powerful way in which gastronomy is made to appear as unprecedented is the use of the historically and geographically specific term "gastronomy" to refer to all cultures of taste. This use makes it seem like the cultivation of taste did not exist before the term "gastronomy" was coined. The universal use of the term transforms the nineteenth-century European version of the human practice of taste cultivation into the first and foremost expression after which all the others are named. Other cultures of taste are thus cast as either earlier but insufficient or later and derivative versions of gastronomy. To avoid this Eurocentric myopia, I reserve the word "gastronomy" to refer exclusively to the regional articulation that the cultivation of taste took in modern Europe, but stripping off its false claim to universality and superiority. Gastronomy is only one particular expression of the many different ways in which humans transform the biological need to eat into culturally and existentially meaningful practices and knowledges.

The narrative of gastronomic progress was formulated and disseminated in print which, together with imperial power, facilitated the imposition of the narrative as incontrovertible. The fact that the narrative was presented in print was subsequently taken as warranting the supposed superiority of gastronomy. According to Priscilla Parkhurst Ferguson, gastronomic writing "signaled the metamorphosis of the consumption of material foodstuffs and corporeal satisfaction into an intellectual and aesthetic pursuit."[35] Statements like this suggest that intellectualization and aestheticization cannot happen without writing. They imply that the taste cultures that predate gastronomy did not approach taste in a thoughtful way and failed to elevate it beyond biological need. This idea is based on the evolutionary view of the history of writing that culminates with alphabetic writing. Jack Goody, one of the first scholars to write about cuisine, was also one of the main advocates of the now discredited "literacy thesis," which claimed that logical thought is dependent on writing.[36] However, Goody revised his position throughout his career, and it is now generally accepted that the uses and consequences of literacy are highly variable: there is no necessary connection between "higher" thinking and the written word.[37] When it comes to culinary cultures, according to Stephen Mennell, the possible consequences of the written word include wider transmission, prescriptive character, faster accumulation and technical cohesion.[38] However, printed texts could have had other effects, and all these effects could also be achieved without the printed word. The printed word is neither a necessity for nor a hindrance to the elaboration of a sophisticated cuisine and taste culture. In the case of gastronomy, the printed word was crucial for the advancement of its authoritarian universalist aspirations.

Scholars of gastronomy continue to build on the distinction between oral and written cultures to establish gastronomy as a more developed kind of culinary culture, arguing that it constituted a field as defined by Pierre Bourdieu. Ferguson maintained that "texts are essential to the intellectualization of food and therefore the constitution of the gastronomic field, whereas a culinary culture incorporates a wide range of representations, most of which will not be intellectualized or even written."[39] Gastronomy could be said to constitute a field as defined by Bourdieu because of its bureaucratization, but it falls short in other defining aspects like having a relative autonomy from economic and political constraints. Gastronomy was fundamentally determined by post-French Revolution politics, capitalism and imperialism. Written texts allowed gastronomy to be intellectualized in a bureaucratized manner, but that does not mean that non-written and non-bureaucratized culinary cultures are not intellectualized. Aside from using a limited definition of intellectualization, the above quoted statement depends on a reductive definition of text limited to alphabetic writing. A more comprehensive concept of text involves "anything that generates meaning through signifying practices."[40] This includes images, sound, objects and indeed activities like cooking. Cultures of taste prior to gastronomy intellectualized taste in many cases without written or printed texts. To appreciate how other cultures of taste intellectualized taste, we need to do more than look for the presence or absence of written texts. We need to learn how to read a wider variety of texts. Alphabetic writing and print are not necessary to intellectualize and aestheticize taste, and they do not necessarily achieve these results either. Gastronomy cannot possibly be the first time that human beings approached taste with a heightened sense of reason and sensibility. This claim only makes sense if we subscribe to a limited understanding of epistemology and aesthetics. There are different ways of defining knowledge, art, beauty and sensuality. Epistemology and aesthetics are only the specific shapes they took in modern Europe. Gastronomy was shaped by the dominant notions of knowledge, art, beauty and sensuality of the context in which it emerged, just like all other taste cultures were shaped by the notions of their own contexts.

The printed word was no doubt a key tool for the establishment of the dominance of gastronomy, but dominance is not the result of excellence. What the printed word achieved for gastronomy was the creation of an authoritarian code that could be easily transmitted and imposed. This code was necessary in a series of processes that are peculiar to the situation in which France found itself in the modern period. One of these processes was the passing down of the taste culture of the French aristocracy to the bourgeoisie after the rather abrupt transition prompted by the French revolution. This is not the first time in history that representatives of a class whose power is waning made use of writing to preserve the taste culture of their dying way of life.[41] Furthermore, in contexts where political transitions have been less abrupt, taste cultures and culinary codes have been successfully passed down and built upon throughout centuries without the need for the written word. This can be illustrated by the many dishes from different parts of the world that have a very long history in which the dish is transformed into different versions in different times and locations but continues to be recognizable.[42] It is significant that when defeated aristocrats and their servant cooks took up the written word to pass their taste culture and culinary knowledge to the bourgeoisie, their main task was to simplify the elaborate recipes of the aristocracy. The written codes of gastronomy were meant to downgrade the sophistication of aristocratic cuisine to create an easy-to-reproduce cuisine that would be marketable to new consumers that did not have the economic or cultural capital to reproduce or appreciate aristocratic cuisine. Gastronomy, as a bourgeois taste culture, was born under an imperative of relative restraint if not mediocrity.

The gastronomic voice in the nineteenth-century French context benefitted from the printed word for the dissemination of a trimmed-down aristocratic taste culture after the ruptures implied by the revolution. In a larger global context, the printed word was an important tool for the subordination of colonized peoples and their taste cultures. Colonialism and imperialism violently disrupted the material conditions of possibility of the taste cultures of colonized peoples. That destruction was accompanied by the use of the printed word as a megaphone for the narrative of gastronomic progress, which silenced the less stridently audible voices of the taste cultures of the colonized. Taste cultures that were not presented in print or that diverged from the gastronomic model in any way were presented as primitive and underdeveloped. Having thus constructed other taste cultures as inferior and in need of development, gastronomers set out to mine them as a source of inspiration. They also took on the task of accumulating knowledge about other taste cultures in an effort to provide the codification that they thought was lacking and indispensable. The process of knowledge accumulation about other taste cultures does not imply a relationship of equality in which the other cultures are understood on their own terms, according to their own concepts of taste, knowledge and art. The narrative of gastronomy used its own voice and perspective to define all cultures of taste and determine their value. The following account by Edmond Neirinck and Jean-Pierre Poulain, in which they make light of the situation, is very telling:

> The whole world turns around France. Given that we should not think that Mongol-style or Chinese-style recipes are dishes of the traditional cuisine of those countries. No way! Hardly a few times one ingredient of the dish finds its origin there. France is at the top of culinary culture, and it is France who rethinks through the light of its science the cuisine of the whole world (On the other hand, this is the time of the great colonial expeditions).[43]

In the nineteenth century, Paris saw itself as the culinary center of France and of the world. The relationship of Paris with the French provinces, the rest of Europe and the rest of the world was one of epistemic domination and disrespect, not one of equality and cosmopolitanism. The relationship between French culinary dominance and colonialism is only mentioned in a parenthesis, as if it were only a curious coincidence not worth more than an aside. This widespread attitude downplays the fact that gastronomy depended materially, epistemically and ideologically on colonialism. While gastronomic scholarship tends to focus on the modernity of gastronomy, it has for the most part failed to account for its coloniality, keeping a veil over their inextricability. The authoritarian voice of the narrative of gastronomy has normalized the Eurocentric perspective on the epistemic and sensorial diversity of the world.

Conflation of Space and Time

The coloniality of the narrative of gastronomic progress is also at work in how it constructs time and space, as the imperial perspective is used to establish what counts as a significant event in the global history of taste. In the narrative of gastronomy, nineteenth-century Paris is firmly established as a major turning point. Words like "revolution," "invention" and "triumph" grace the titles of books devoted to the study of French gastronomy.[44] The act of presenting nineteenth-century Paris as the starting point of a new chapter in the history of taste cultures was simultaneously an act of marginalizing other taste cultures. The specific developments of Europe are taken as the teleological destiny for all, turning differences into earlier stages of a universal development. Whereas France developed modern gastronomy, other culinary knowledge systems are reduced to being just "traditional" or "ethnic." Whereas gastronomy is presented as relatively autonomous if not altogether free, all other taste cultures are portrayed as fixed in a past devalued as pre-modern, and captive of the limitations of specific places and cultures. In the gastronomic understanding of space and time, the cultures of taste from places other than Europe become cultures of earlier times and earlier developmental stages. Therefore, they are not just different and located elsewhere but also inferior and left behind in time by gastronomy. The narrative of gastronomy thus follows the familiar historicist structure of global historical time that Chakrabarty summarized as "first in Europe, then elsewhere."[45]

The narrative of gastronomy develops through turning points in a path of linear progress in which the West is constructed as the leader while other parts of the world are portrayed as just catching up. Reductive as this narrative is, it still underlies most accounts of the global diversity of cultures of taste. Global historical accounts of gastronomy usually grant, if only in passing, that other sophisticated taste cultures existed outside of Europe before gastronomy. It is also often argued that other taste cultures have their own "alternative modernities."[46] However, these gestures of inclusion do not alter the master narrative or the power relationship. Granting that other taste cultures are modern in alternative ways still leaves gastronomy as the privileged reference point, while the others remain marginal as just alternatives to the legitimate one.[47]

The turning of different geographical places into different time periods can be observed in many books proposed as global historical accounts of food, cuisine and taste. The chronologically arranged chapters of such books include places like China and the Islamic world only in the chapters dedicated to periods framed as "pre-modern."[48] Once the chapters reach the modern period, the focus is almost exclusively on the Western world, leaving other regions firmly tucked in the past. Different geographical spaces are turned into time periods when Greece and Rome appear in the narrative only in Antiquity, the Middle East is brought up only in the Middle Ages and China disappears after the Imperial era. Latin America tends to be mentioned only briefly after Columbus' arrival, while Africa is almost never mentioned, and the United States enters mostly in the high and late capitalism period. Europe is the only place shown to have a continuous historical presence. The historical narrative of gastronomy replicates and naturalizes the contingent history of global geopolitics. The Hegelian presupposition here is that the history of cuisine moves from east to west, and the west is the end of culinary history. In this relay race view of history, once the West enters the field of sophisticated taste cultures it becomes the sole protagonist of their universal history. Other parts of the world presumably are either frozen in time, decayed, or trying to catch up, and therefore unworthy of attention. In these books, the inclusion of cultures of taste other than gastronomy serves as background to highlight the rise of gastronomy as the ineludible end point in the history of taste.

The accounts of the history of taste that take gastronomy as its end point transform taste preferences and culinary changes in Europe into stages in a universal evolutionary history. Changes in taste preferences are not explained as contingent to their specific contexts but as steps forward in the march of progress. The most important example of this is the shift away from the use of spices in modern Europe, which is more fully examined in Chapter 4. The modern culinary identity of Europe and its sense of cultural superiority were founded on this shift. The peoples who "still" cook with spices are portrayed as stuck on the cooking style of yesteryear. Even minor changes in taste preferences and culinary techniques are presented as steps forward. In the history of sauce making, for example, the use of butter as opposed to lard or oil is dubbed "one of the most important culinary transformations."[49] The fact that in India *ghee* (clarified butter) had been in use for centuries is either unknown or deemed irrelevant.[50] Culinary changes are furthermore presented as necessary and pre-ordained, leading to the final consummation of gastronomic ideals. This logic transforms preparations like *jus* and *coulis* into the "forefathers of our fonds,"[51] and Taillevent into a visionary who "would seem to prefigure the great Antonin Carême" because he sieved a cameline sauce.[52] The historical data of course does not fit neatly into this narrative, as was noted by Jean-François Revel without a shred of irony: "Should duck with oysters (yet another creation of Massialot's) be considered a holdover from the Middle Ages or as a forebear of a modern dish? There is no satisfactory answer."[53] Revel confesses perplexity when he finds characteristics that he considers to be modern in "pre-modern" cooking, yet he does not realize that this is only perplexing if he presupposes the linearity of history. The changes in culinary ideals are furthermore presented as moral progress, as if the preferences of other times and places were not only less developed but also lacking in morality. This is suggested by casual judgments that appear in gastronomic historiography. The moral superiority of modern taste preferences is presumed when scholars describe medieval European cuisine as combining sweet and savory flavors "promiscuously,"[54] and as indulging in a "spice-orgy."[55] Statements like these demonstrate that both gastronomy and much of its scholarship wear the blinders of Euro-centered Christianity.

Western Protagonism

The main character and sovereign agent in the standard narrative of the global history of food and taste is the West. The West is portrayed both as the only agent with the capacity of affecting the history of taste and as utterly unaffected by the actions of others. Cultures of taste other than gastronomy are given considerably less attention, and many are not even mentioned. This is the result of equating the capacity to spread globally with a higher level of intrinsic value. The peoples of the regions of the world that the narrative of gastronomy constructs as frozen in the past are represented as having flaws and limitations that justify their subordinate status. Nobody has expressed this more openly than Christian Boudan in his book on the geopolitics of taste. He excluded African cuisine from his book because it "did not undertake any expansion outside of its zone, and continues to be virtually unknown outside of the continent."[56] This is a violent erasure of the culinary agency and creativity of Africans and of the African diaspora. Enslaved Africans were responsible for the spread of African culinary and agricultural techniques, and ultimately for the creation of new cuisines, most notably in the greater Caribbean region.[57] The supposed lack of spread of West African cuisines, according to Boudan, is due to their being "characterized by the deficit of fresh proteins and the traditional use of sun-dried foods," which makes them unattractive to peoples with more resources.[58] He presupposes that the cuisine of people with fewer resources cannot possibly produce anything worthy of mention in the history of cultures of taste. The marginalization that the African continent has been subjected to since the dawn of the modern era is reinforced by the negation of African peoples' ability to create cultures of taste that have more value than just assuring subsistence. The negation of the creativity and cultural agency of the poor or less powerful continues today with the lack of recognition of the fundamental role of immigrant restaurateurs in the shaping of cosmopolitan taste, as has been argued by Krishnendu Ray.[59]

In the narrative of gastronomic progress, the agency of non-European peoples is downplayed or even erased. An obvious example of this is how Arabic-speaking peoples are presented as transporters of goods and culture, rather than as active participants in the creation and transformation of a shared Afro-Eurasian culture. This narrative strategy allowed Europe to construct itself as not being related, and much less owing anything, to the "Orient," which it constructed as its other. It also allowed Europe to create the illusion of a pure cultural continuity between modern Europe and classical antiquity. In the modern era, characterized by European imperialism and colonialism, colonized and impoverished peoples are frequently represented as totally vanquished and as passive receivers of the culture of the colonizers. An example of this is seen in how India is presented in global histories of taste. The lack of willingness to grant India any measure of cultural agency is remarkable. Jack Goody, for example, has stated: "in North India, the rulers imported the practices of Islamic culture, later modified by the advent of the British."[60] In this statement, India is credited only with importing Islamic cuisine, apparently without modifying it all. The ability of making transformations is only given to the British colonizers. This perspective is constantly repeated even though Mughlai cuisine, as developed in India, is more complex than what was brought in by the Mughals and more sophisticated than the simplified version popularized by the British.

Another way in which the narrative of gastronomy negates the agency of people other than modern Europeans is by presenting them as captive of their religious and medical thought. The narrative presents taste cultures other than gastronomy as fully determined by religion and medicine. All religions tend to impose food restrictions and generally aim to regulate pleasure. Religious discourses indeed impose a specific kind of discipline on the act of eating, but they also add meaning to the consumption of many foods, which in many cases elevates the experience of taste. There is no reason to believe that religion is incompatible with a sophisticated culture of taste with an emphasis on pleasure. All taste cultures are shaped by a number of discourses that impose limits. Religion, medicine, rationalism, aesthetics and nationalism are only a few examples. The problem with the modern perspective is that it fails to recognize rationalism and aesthetics as regulating and contingent discourses and naively asserts that gastronomy is free from extraneous constraints.

There is a tendency to see culture in societies where religion has a central role as static or monolithic. The taste cultures of India and of the Islamic world are quickly discounted in global histories of taste on account of their relationship to Hindu and Islamic thought. In the narrative of gastronomy, Islamic and Hindu cultures are presented as homogenous and unchanging, and as incapable of developing a pleasure-oriented approach to food. These views present Hindus and Muslims as paralyzed prisoners of their cultures without any agency or control over them. The actual existence of sophisticated taste cultures in both Hindu and Muslim contexts clearly contradicts these views, but the template for analysis provided by the narrative of modernity obscures their existence.

Medieval European culinary cultures have also been treated with disdain from the gastronomic perspective, but medieval Europeans are usually granted a little more agency to negotiate and transform religious mandates. Boudan, for example, praises European creativity in the face of the Christian fasting requirements arguing that they compensated for Lent with "the creation of an exceptional fish cuisine."[61] There should be an equal appreciation of the Eid feasts that follow the fast of the Ramadan and of the elaborate vegetarian cuisines of India. Indeed, one of the most admired cuisines these days is the one produced by ascetic Buddhist monks in Japan. This is a clear case in which religious discipline elevated a cuisine to high levels of sophistication and transcendence.

The modern perspective also puts a lot of stress on the religious proscription of beef in the case of Hindus and of pork in the case of Jews and Muslims. The centrality of meat in modern Western diets makes it hard to understand that the lack of consumption of beef or pork does not inhibit the formation of a sophisticated taste culture any more than not eating dogs, cats, insects or any other specific food does. Looking down on taste cultures that do not use beef or pork only exposes the provincial character of the gastronomic perspective. Ironically, while all religions impose restrictions that taste cultures negotiate or even use to their advantage, there is one clear way in which Christianity presented a more serious obstacle for the enjoyment and development of taste than other religions. Christian thought directly imposes restraint in the enjoyment of all food by warning against eating in excess or with too much zeal. This imposition directly interfered with the pleasures of taste and yet the narrative of gastronomy downplays its impact while exaggerating the significance of the ban against eating beef or pork.

The narrative of gastronomy also puts down any taste culture that is articulated with a medical understanding of food and diet. This is based on the false presupposition that gastronomy was free from medical influences. This is another example of how the narrative of gastronomy posited its own characteristics as the most advanced. The relationship between gastronomy and modern understandings of medicine is not considered to interfere with the pleasures of taste, but any other understanding of medicine is ruled as an obscurantist impediment. As Jean Louis Flandrin informs us:

> The idea that the function of cooking is to make food easy to digest, and that culinary creation takes place within a system of dietetic constraints whose purpose is to balance hot against cold and dry against moist, will seem obvious today to anyone from India, China, or the West Indies. Yet Europeans find it difficult to admit that the same idea once shaped their own attitudes toward food.[62]

In Flandrin's narrative, places as diverse as China, India and the West Indies are lumped together because their culinary cultures refer to dietary systems. According to Flandrin, Europeans moved away from dietetic concerns in their gastronomic cooking so fully that they cannot even understand that they once shared that kind of thinking with the rest of humanity. If it is difficult to admit this, it is because the European sense of identity and superiority is based on dubious differences like this one.

Modern medical beliefs did not present any substantial advantage compared to other medical systems when it comes to freedom to cultivate taste. Many of the dishes designed according to the humoral medical system continue to be relished today, even when cooks and eaters might have no knowledge of the system. The humoral dietary system, based on Arab, Jewish and Greek sources, dominated medical thought in Europe until the middle of the seventeenth century. This system directly influenced cooking, requiring the use of condiments to balance the humors of each food and to correct the humor imbalances of the eaters. For example, melon would be served with cheese or salted meats because saltiness was expected to correct the excessive and dangerous wetness of melons.[63] The fact that we continue to enjoy melon with prosciutto suggests that gustatory concerns were not foreign to humorology-influenced cooking. As explained by Ken Albala, medicine and cuisine were engaged in a conversation, but their relationship was ambiguous, as cooks did not always obey the mandates of medicine even before the two fields drifted apart.[64] Similarly, there is an extensive culinary repertoire in India originally inspired by Ayurvedic principles that is considered delectable by eaters who have no understanding of or interest in Ayurveda. Clearly, the people who developed recipes in humoral and Ayurvedic contexts were not blindly following medical advice and always kept gustatory interests in mind. The systems were elastic enough to accommodate change and pleasure.

The other side of the portrayal of people other than modern Europeans as lacking agency to create taste-oriented culinary cultures is the representation of modern Europeans as the ones with agency to define and transform their own and all other culinary cultures. The narrative of gastronomy and gastronomic scholarship are distinguished by a remarkably inadequate understanding of the complex processes of transculturation, localization and hybridization that follow any kind of cultural encounter. Contemporary cultural criticism, in fields like Latin American and post-colonial studies, generally agrees on the basic notion that cultural encounters, no matter how unequal the power relationship, always involve multidirectional cultural transformations. Cultural transformation includes the creation of new forms of culture, not just mixing previously existing ones. Colonialism transforms both the culture of the colonized and the culture of the colonizers, and both colonizers and colonized are agents in the process of transformation, even if in different ways. In the narrative of gastronomy, however, cultural transmission in colonial contexts is presented as unidirectional. Colonized peoples are presented as accepting the culture of the colonizers without creative transformations, while the colonizers are portrayed either as unchanged by colonialism or as in total control of the changes.[65] This is most evident in the treatment of the biological and cultural exchanges unleashed by the European colonization of the Americas, which is euphemistically called the "Columbian Exchange."

The lack of understanding of the complexities of cultural encounters has made it possible for the "Columbian Exchange" to be presented as having no effect on European taste culture. Massimo Montanari, for example, defends the agency of Europeans in their incorporation of foods from the Americas: "their acceptance was made possible only by a process of cultural confirmation which changed, sometimes radically, their utilization in such a way as to adapt them to specifically local traditions."[66] The narrative of gastronomy congratulates Europeans for their ability to adapt the foods that came from elsewhere, as if the same amply demonstrated human ability did not apply to Amerindians and everyone else. Culture is just never transmitted wholesale or unilaterally. European food cultures took hold on the Americas not because Amerindians were necessarily eager to do so, as if European foods were universally more desirable, but because colonization involved the destruction of their food systems and the imposition of new ones. In spite of the abysmally asymmetrical power relations between Amerindians and Europeans, the indigenous populations of the Americas (and eventually Africans, Asians and Creoles) transformed the culinary ingredients, techniques and ideas that came from Europe and beyond, ultimately creating new cuisines and taste cultures.

The process of culinary transculturation and hybridization that took place in the Americas after colonization is recognized at least at some level by most scholars. The colonization of the Americas was the base on which modern European affluence and power were built. It was also a major event from a global biological and cultural perspective.[67] However, Rachel Laudan has claimed that the Columbian Exchange was "a largely one-way culinary transfer,"[68] meaning that Europeans did not adopt the culinary knowledge of the Americas. Whereas it is true that there are few straightforward culinary transfers from the Americas to Europe, that does not mean that Amerindian culinary cultures had no impact on European taste culture. Amerindian culinary cultures impacted European ones for centuries because Europeans were simultaneously attracted to and fearful of Amerindian food culture.

The apparent lack of impact of Amerindian culinary knowledges and techniques on European culinary cultures is the result of the apprehensions of the colonizers, who were afraid of losing their ill-gained position of global power. Rebecca Earle has discussed at length the anxiety that the Spanish felt regarding the foods of the Americas.[69] According to humoral medical theories, bodies are always in flux, so eating the food of Amerindians could have transformed them into Amerindians.[70] Given that the colonial enterprise was in large part based on the idea of the superiority of European Christian bodies, eating European foods was essential for keeping colonial hierarchies. However, the Spanish conquerors couldn't possibly just transplant their culinary practices in the Americas while remaining unaffected by local food and taste. Even though Europeans resisted changing their diets, their material dependence on the colonized made them learn to process and taste foods according to indigenous patterns. Marcy Norton has explained how the incorporation of tobacco and chocolate into European culture included not only the knowledge of cultivation of the raw materials but also the organoleptic ways of enjoying them and the social behaviors surrounding their consumption.[71] Norton argues that people in Europe adopted tobacco and chocolate not in spite of the meanings that Amerindians gave to them but because of them.[72] The examples of tobacco and chocolate clearly show that Amerindian food culture and aesthetic values, not just ingredients as raw materials, had a deep impact in Europe.

A similar resistance to the food culture of the colonized was seen in the case of British colonizers in India in the nineteenth century,[73] even though by then the closed, essentialized and racialized notion of the body had taken the place of the open notion of the body found in humorology. This resistance betrays the insecurity of colonizers, who themselves didn't seem to fully believe in the fantasies of natural white Christian superiority.[74] In any case, anxiety and resistance could not possibly stop the multidirectional impacts that characterize all cultural contacts. Even today, the embracing of Indian food by the British continues to be thoroughly mediated by racist apprehensions and fantasies.

The foods and culinary knowledge that Europeans were too fearful or disdainful to adopt had at least as much impact on their cuisines as the foods and customs that they did adopt. Modern race discourse relied on food to establish identities and hierarchies. For Europeans to hold on to their supposed racial superiority they had to accept significant limitations to what and how they could eat. The need to differentiate themselves from colonized peoples led them to close their culinary culture to affect and to establish a taboo regarding foods that they considered were most representative of the peoples they constructed as inferior races. Spices and chiles were racialized and proscribed from a civilized and refined diet. This taboo against affective flavors represents for gastronomy a limitation more serious than any other, given that it really curtails the range of possibilities of the experience of taste.

The resistance to learning from the culinary knowledge of colonized peoples is constitutive of modern taste. Europe was committed to maintaining a limited and provincial culinary culture in order to preserve its fragile sense of difference and superiority. The colonization of the Americas changed European culinary culture by making them fearful of change, and unable to enjoy highly affective foods. These foods were ruled as deviant, and their eventual and partial incorporation was possible only as a transgressive pleasure. Gastronomy was built around many significant exclusions and anxieties that ultimately made modern European taste culture extremely limited and provincial in spite of its pretensions to cosmopolitanism and universality. Race, as discussed in Chapter 4, is one of the structural pillars of modern taste. The narrative of gastronomy does not innocently celebrate a culinary culture that is dear to many; it continues to naturalize and reinforce modern/colonial racial hierarchies.

A Borgesian Parody

A critique of the narrative of gastronomic progress can be found in the short story "An Abstract Art," written in 1967 by Argentine writers Jorge Luis Borges and Adolfo Bioy-Casares.[75] This story, like all stories co-authored by Borges and Bioy-Casares using the narrative voice of the fictional cultural commentator Bustos Domecq, is humorous. "An Abstract Art" takes the shape of a historical narrative about the origins of a very special type of restaurant called a tenebrarium. This story is usually read as a satire of the pieties of modernist movements. The tenebrarium is the culmination of many attempts to create a "culinary cooking" that owed nothing to the plastic arts or to the alimentary purpose.[76] The narrator informs us that this ideal was formulated by the high priest of pure cuisine, Pierre Moulonguet, in the *Manuel Raissoné* which is a three-volume manifesto written in verse. The doctrinarian fervor of the characters in the story should make readers laugh at the scientific pretensions and dogmatism of modernism. The ideal of a pure cuisine echoes the actual modern gastronomic dream of liberating taste from non-gustatory influences. However, the development of the story exposes the absurdity of the gastronomic liberation ideal and undermines modern progressivist historical narratives in general. Readers already familiar with the chronicles of Bustos Domecq know that the narrator is a pompous but mediocre provincial newspaper cultural commentator.[77] Borges commented that his writing with Bioy-Casares always took the form of a humor competition.[78] However, readers of "An Abstract Art" do not need extra-textual clues to realize the humorous nature of the story. The story ridicules the reverent tone of the standard historical narratives that the narrator Bustos Domecq clumsily imitates. The story's construction of space, voice, time and characters is akin to the ones used in the narrative of gastronomic progress discussed above.

The story's point of departure is the modern construction of space as time, in which the West is ahead, and the rest of the world is behind. The opening sentence of the story accepts and laments the presumed backward status of the Argentine cultural space: "[…] the fact must be faced that at this late date a tourist Mecca of the modern New World like Buenos Aires boasts but a single *tenebrarium* […]."[79] The chronicle of the invention of the tenebrarium is told by Bustos Domecq using an authoritative voice that wants to show off his knowledge as evidence of high culture credentials in spite of his provincial Argentine origin. The peripheral nature of

Argentina, destined to be a pale imitator of imported modern progress, is invoked to set the tone of the deferential attitude with which presumably Argentine readers should read the chronicle. However, throughout the story the narrator's exaggerated reverence allows readers to appreciate humorous aspects of the chronicle, even though the narrator does not seem aware of them. Readers laugh both at the contents of the chronicle and at the narrator, who takes too seriously a historical narrative that is clearly laughable. The voice of Bustos Domecq is a parody of the unwarranted self-assurance of the Eurocentric voice of modern historiography.

The chronicle of the invention of the tenebrarium takes the structure of the standard narrative of modern progress led by the West and reveals it as comically preposterous. The development of the evolutionary history that led to the creation of the tenebrarium goes through a series of stages spurred by European luminaries. The pomposity of Bustos Domecq contrasts with the underwhelming formulations and discoveries made by the protagonists of his account. One of the intellectual ancestors of the tenebrarium is a character called Frans Praetorius, who in a manual called "Les Saveurs" established that there are four fundamental flavors: acid, salty, insipid, and bitter. Bustos Domecq reports:

> In 1891, Praetorius publishes the today classical *Les Saveurs*; let us not forget, by the way, that the Grand Man, yielding with unimpeachable good will to a host of unknown correspondents, adds to his previous catalog a fifth taste, that of sweetness, which, for reasons it would be impertinent to go into here, had hitherto eluded his perspicacity.[80]

The author of the manual had scientific pretensions and is venerated even though he failed to account for the generally undisputed fact that sweetness is a basic flavor. Bustos Domecq acknowledges Praetorius' oversight, but this does not make him question the validity of his narrative or change his reverent tone. Praetorius' only real achievement was to write down something obvious and not even in a fully correct way. The absurdity of admiring Praetorious for such an underwhelming feat is reminiscent of many of the so-called achievements of gastronomic writing and of the fetishization of the printed word.

The stages that led to the invention of the tenebrarium begin with a restaurant in which diners could choose between "a lump of sugar, a cube of aloes, a cotton wafer, a grapefruit rind, or a *granum salis*."[81] This restaurant was followed by another one that served pyramids of pure flavors but failed after they began serving pyramids of mixed flavors, which compromised the ideal of purity. Another ancestor of the tenebrarium is a restaurant that served classic dishes but half-liquefied, reduced to a gray mucilaginous mass. All these restaurants, following the modernist logic, were intended to allow for the appreciation of taste in a pure state but were ultimately unappealing. In the end the tenebrarium, the restaurant that according to the narrative achieved the ideal of allowing customers to focus only on taste, was invented by a regular restaurateur in Geneva who served the usual foods, but with the lights turned off. The absurdity of the idea is beside the point for Bustos Domecq, for whom the modernist ideal of the pure experience of taste is more important than having a pleasant experience of taste.

Bustos Domecq narrates the invention of the tenebrarium as if it were a major milestone in the history of humanity. The silliness of dining in the dark as a way of focusing on taste exclusively is stressed by the fact that what Bustos Domecq remembers from dining at the tenebrarium is the metallic music of the cutlery and the occasional crash of a breaking tumbler.[82] Instead of providing a pure taste experience, the tenebrarium left a chaotic auditory impression. The humor of Borges and Bioy-Casares' short story is produced by the narrator's lack of awareness of how inconsequential the history of the tenebrarium really is. The narrative of gastronomy is in many ways similar to the narrative of the tenebrarium. The mystique and power of the narrative of gastronomy as if it were the first and foremost culture of taste continues to have a hold on the popular and academic imagination even when its lack of validity is not hard to see.

Notes

1 Whereas France took the lead in the development of gastronomy, it did so in the context of more widespread transformations in Europe. The particular reasons why France, rather than England, took the lead have been analyzed by Stephen Mennell, *All Manners of Food: Eating and Taste in England and France from the Middle Ages to the Present*, 2nd ed. (Urbana, IL: University of Illinois Press, 1996). Contrary to explanations based on ideas of French exceptionalism, Mennell found that the French court provided the model for all of Europe because French nobility, lacking substantial military and governmental power, centered its attention on social display. My own analysis is concerned with the global modern/colonial articulation of gastronomy, more than with the specific roles of different European countries.

2 Hayden V. White, *Metahistory: The Historical Imagination in Nineteenth-Century Europe* (Baltimore, MD: Johns Hopkins University Press, 1973), 6–7.

3 Aníbal Quijano, "Coloniality and Modernity/Rationality," *Cultural Studies* 21, no. 2 (March 2007): 174, https://doi.org/10.1080/09502380601164353.

4 Dipesh Chakrabarty, *Provincializing Europe: Postcolonial Thought and Historical Difference* (New Delhi: Oxford University Press, 2000), 27.

5 Jean Bottéro, *Textes culinaires mésopotamiens/Mesopotamian culinary texts* (Winona Lake, IN: Eisenbrauns, 1995), 7.

6 H.D. Miller, "The Pleasures of Consumption: The Birth of Medieval Islamic Cuisine," in *Food: The History of Taste*, ed. Paul H. Freedman (Berkeley: University of California Press, 2007), 140.

7 Miller, 144–145.

8 Joanna Waley-Cohen, "The Quest for Perfect Balance: Taste and Gastronomy in Imperial China," in *Food: The History of Taste*, ed. Paul Freedman (Berkeley: University of California Press, 2007), 101.

9 E.N. Anderson, *The Food of China* (New Haven, CT: Yale University Press, 1988), 69–93.

10 Joanna Waley-Cohen, "The Quest for Perfect Balance: Taste and Gastronomy in Imperial China," 114–115.

11 Waley-Cohen, 128–129.

12 K.T. Achaya, *Indian Food: A Historical Companion* (Delhi: Oxford University Press, 1998), 45.

13 Achaya, 45–46. The āppam is a rice pancake baked on a concave clay vessel and the idi-āppam is made by extruding a dough of boiled and mashed rice to form

thread-like noodles. The dōsai and the adai are both pan fried, but the dōsai is made with a batter of fermented rice and urad dhāl (black gram), while the adai batter is a mixture of rice and a variety of pulses.

14 Achaya, 77.
15 Achaya, 79.
16 Achaya, 54.
17 Achaya, 54.
18 Achaya, 118.
19 Achaya, 119.
20 Abul Fazl 'Allami, and H. Blochmann, *Ain I Akbari*, vol. 1 (Calcutta: Asiatic Society of Bengal, 1873), 56–73.
21 Abul Fazl 'Allami and Blochmann, 73–87.
22 Bernardino de Sahagún, *Historia general de las cosas de Nueva España*, ed. Juan Carlos Temprano (Madrid: Dastin historia, 2001).
23 Sahagún, 658–663.
24 Sahagún, 798–799.
25 Tadeusz Lewicki, *West African Food in the Middle Ages: According to Arabic Sources* (Cambridge: Cambridge University Press, 2008).
26 Jessica B. Harris, *High on the Hog: A Culinary Journey from Africa to America* (New York: Bloomsbury, 2011), 11–14.
27 Paul Freedman, "Food Histories of the Middle Ages," in *Writing Food History: A Global Perspective*, ed. Kyri Claflin and Peter Scholliers (London; New York: Berg, 2012), 25.
28 Walter Mignolo, "The Enduring Enchantment (or The Epistemic Privilege of Modernity and Where to Go from Here)," *The South Atlantic Quarterly* 101, no. 4 (2002): 938.
29 James Beard Foundation, "The James Beard Foundation's Mission Is to Celebrate, Nurture, and Honor America's Diverse Culinary Heritage through Programs That Educate and Inspire." http://www.jamesbeard.org/about. Accessed September 1, 2014. The stated mission of the foundation has since changed.
30 Mitchell Davis and Anne McBride, "The State of American Cuisine: A White Paper Issued by the James Beard Foundation Based on Surveys Conducted as Part of the 2007 James Beard Foundation's Taste America® National Food Festival." (James Beard Foundation, July 2008), 11.
31 Jeffrey Pilcher, *Food in World History* (New York: Routledge, 2006), 34.
32 Pilcher, 41.
33 Santiago Castro-Gómez, "The Missing Chapter of Empire: Postmodern Reorganization of Coloniality and Post-Fordist Capitalism," *Cultural Studies* 21, no. 2 (March 2007): 433, https://doi.org/10.1080/09502380601162639.
34 Fernando Coronil, "Beyond Occidentalism: Toward Nonimperial Geohistorical Categories," *Cultural Anthropology* 11, no. 1 (February 1996): 57.
35 Priscilla Parkhurst Ferguson, *Accounting for Taste: The Triumph of French Cuisine* (Chicago: University of Chicago Press, 2004), 11.
36 Jack Goody and Ian Watt, "The Consequences of Literacy," *Comparative Studies in Society and History* 5, no. 3 (April 1963): 304–345.
37 John Halverson, "Goody and the Implosion of the Literacy Thesis," *Man, New Series* 27, no. 2 (June 1992): 301–317.
38 Stephen Mennell, *All Manners of Food*, 67.
39 Ferguson, 107.
40 Chris Barker, *The Sage Dictionary of Cultural Studies* (London: Sage Publications, 2004), 199.
41 One notable example is the memoir of the late Ming dynasty historian Zhang Dai. For a discussion, see Chapter 5.
42 One example is nicely documented in Nawal Nasrallah, "In the Beginning There Was No Musakka: A Curious Case in the History of Culinary Metamorphoses," *Food, Culture & Society* 13, no. 4 (December 2010): 595–606.

43 Edmond Neirinck and Jean-Pierre Poulain, *Historia de la cocina y de los cocineros: Técnicas culinarias y prácticas de mesa en Francia, de la Edad Media a nuestros días* (Barcelona: Editorial Zendrera Zariquiey, 2001), 73. My translation.

44 See Susan Pinkard, *A Revolution in Taste: The Rise of French Cuisine, 1650–1800* (Cambridge: Cambridge University Press, 2009); Amy B. Trubek, *Haute Cuisine: How the French Invented the Culinary Profession* (Philadelphia, PA: University of Pennsylvania Press, 2001); and Priscilla Parkhurst Ferguson, *Accounting for Taste: The Triumph of French Cuisine* (Chicago: University of Chicago Press, 2004).

45 Chakrabarty, 7–8.

46 The idea of alternative modernities was explored in Dilip Parameshwar Gaonkar, ed., *Alternative Modernities* (Durham, NC: Duke University Press, 2001).

47 For a discussion of the importance of looking for alternatives to modernity as opposed to alternative modernities, see Walter Mignolo, *The Darker Side of Western Modernity: Global Futures, Decolonial Options* (Durham, NC: Duke University Press, 2011), xxvii–xxxi.

48 See for example Reay Tannahill, *Food in History* (New York: Crown Trade Paperbacks, 1995); Jean Louis Flandrin and Massimo Montanari, *Food: A Culinary History from Antiquity to the Present*, ed. Albert Sonnenfeld (New York: Penguin Putnam, 2000); Paul H. Freedman, ed., *Food: The History of Taste* (Berkeley: University of California Press, 2007).

49 Christian Boudan, *Geopolítica del gusto: La guerra culinaria*, trans. Marie-Anne Salaün (Gijón: Trea, 2008), 343. All translations from Spanish to English are my own.

50 Ghee is mentioned in the literary work of the Aryans, which spans from 1500 to 350 BCE. Achaya, *Indian Food*, 33.

51 Neirinck and Poulain, 40.

52 Jean-François Revel, *Culture and Cuisine: A Journey through the History of Food* (Garden City, NY: Doubleday, 1982), 102.

53 Revel, 164.

54 Pilcher, 36.

55 Fernand Braudel, *Civilization and Capitalism, 15th–18th Century* (New York: Harper & Row, 1982), 221.

56 Boudan, 402.

57 For details see Harris.

58 Boudan, 402.

59 Krishnendu Ray, *The Ethnic Restaurateur* (London: Bloomsbury Academic, 2016).

60 Jack Goody, *Cooking, Cuisine and Class: A Study in Comparative Sociology* (Cambridge: Cambridge University Press, 1996), 127.

61 Boudan, 327.

62 Flandrin and Montanari, 418.

63 Ken Albala, *Eating Right in the Renaissance* (Berkeley: University of California Press, 2002), 254–255.

64 Albala, 241–283.

65 The field of postcolonial studies has produced many volumes that analyze the complexities of food- and taste-related interactions in different colonial contexts. These volumes should affect how the global history of food and taste is told.

66 Massimo Montanari, *The Culture of Food*, The Making of Europe (Oxford, UK; Cambridge, MA: Blackwell, 1996), 132.

67 Alfred W. Crosby, *The Columbian Exchange: Biological and Cultural Consequences of 1492* (Westport CT: Greenwood, 1972).

68 Rachel Laudan, *Cuisine and Empire: Cooking in World History* (Berkeley, CA: University of California Press, 2013), loc. 4481 of 13660, Kindle.

69 Rebecca Earle, *The Body of the Conquistador: Food, Race, and the Colonial Experience in Spanish America, 1492–1700* (Cambridge: Cambridge University Press, 2013).

70 Earle, 2–3.

71 Marcy Norton, *Sacred Gifts, Profane, Pleasures: A History of Tobacco and Chocolate in the Atlantic World* (Ithaca, NY: Cornell University Press, 2008).

72 Norton, 2.
73 E.M. Collingham, *Imperial Bodies: The Physical Experience of the Raj, c. 1800–1947* (Cambridge, UK; Malden, MA: Polity Press; Blackwell Publishers, 2001).
74 Earle, 214.
75 Jorge Luis Borges and Adolfo Bioy-Casares, "An Abstract Art," in *A Literary Feast: An Anthology*, ed. Lilly Golden (New York: Atlantic Monthly Press, 1995), 70–73.
76 Borges and Bioy-Casares 73.
77 Emir Rodríguez Monegal, "Mundo Nuevo - Nota Sobre Biorges," October 17, 2007, https://web.archive.org/web/20071017043300/http://www.archivodeprensa.edu. uy/r_monegal/bibliografia/prensa/artpren/mundo/mundo_22.htm. Accessed June, 2022.
78 Mireya Camurati, "Las 'Crónicas de Bustos Domecq' y la subversión de la realidad," *Modern Language Studies* 15, no. 1 (1985): 32, https://doi.org/10.2307/3194415.
79 Borges and Bioy-Casares, "An Abstract Art," 70.
80 Borges and Bioy-Casares, 71.
81 Borges and Bioy-Casares, 71.
82 Borges and Bioy-Casares, 70.

References

Allami, Abul Fazl, and H. Blochmann. *Ain I Akbari*. Vol. 1. Calcutta: Asiatic Society of Bengal, 1873.

Achaya, K.T. *Indian Food: A Historical Companion*. Delhi: Oxford University Press, 1998.

Albala, Ken. *Eating Right in the Renaissance*. Berkeley: University of California Press, 2002.

Anderson, E.N. *The Food of China*. New Haven: Yale University Press, 1988.

Barker, Chris. *The Sage Dictionary of Cultural Studies*. London: Sage Publications, 2004.

Borges, Jorge Luis, and Adolfo Bioy-Casares. "An Abstract Art." In *A Literary Feast: An Anthology*. Edited by Lilly Golden, 70–73. New York: Atlantic Monthly Press, 1995.

Bottéro, Jean. *Textes culinaires mésopotamiens/Mesopotamian culinary texts*. Winona Lake, IN: Eisenbrauns, 1995.

Boudan, Christian. *Geopolítica del gusto: La guerra culinaria*. Translated by Marie-Anne Salaün. Gijón: Trea, 2008.

Braudel, Fernand. *Civilization and Capitalism, 15th–18th Century*. New York: Harper & Row, 1982.

Camurati, Mireya. "Las 'Crónicas de Bustos Domecq' y la subversión de la realidad." *Modern Language Studies* 15, no. 1 (1985): 30. https://doi.org/10.2307/3194415.

Castro-Gómez, Santiago. "The Missing Chapter of Empire: Postmodern Reorganization of Coloniality and Post-Fordist Capitalism." *Cultural Studies* 21, no. 2 (March 2007): 428–448. https://doi.org/10.1080/09502380601162639.

Chakrabarty, Dipesh. *Provincializing Europe: Postcolonial Thought and Historical Difference*. New Delhi: Oxford University Press, 2000.

Collingham, E.M. *Imperial Bodies: The Physical Experience of the Raj, c. 1800–1947*. Cambridge, UK; Malden, MA: Polity Press; Blackwell Publishers, 2001.

Coronil, Fernando. "Beyond Occidentalism: Toward Nonimperial Geohistorical Categories." *Cultural Anthropology* 11, no. 1 (February 1996): 51–87.

Crosby, Alfred W. *The Columbian Exchange: Biological and Cultural Consequences of 1492*. Westport, CT: Greenwood, 1972.

Davis, Mitchell, and Anne McBride. "The State of American Cuisine: A White Paper Issued by the James Beard Foundation Based on Surveys Conducted as Part of the 2007 James Beard Foundation's Taste America® National Food Festival." James Beard Foundation, July 2008.

Earle, Rebecca. *The Body of the Conquistador: Food, Race, and the Colonial Experience in Spanish America, 1492–1700*. Cambridge: Cambridge University Press, 2013.

Ferguson, Priscilla Parkhurst. *Accounting for Taste: The Triumph of French Cuisine*. Chicago: University of Chicago Press, 2004.

Flandrin, Jean Louis, and Massimo Montanari. *Food: A Culinary History from Antiquity to the Present*. Edited by Albert Sonnenfeld. New York: Penguin Putnam, 2000.

Freedman, Paul, ed. *Food: The History of Taste*. Berkeley: University of California Press, 2007.

———. "Food Histories of the Middle Ages." In *Writing Food History: A Global Perspective*. Edited by Kyri Claflin and Peter Scholliers, 24–37. London: Berg, 2012.

Gaonkar, Dilip Parameshwar, ed. *Alternative Modernities*. Durham, NC: Duke University Press, 2001.

Goody, Jack. *Cooking, Cuisine and Class: A Study in Comparative Sociology*. Cambridge: Cambridge University Press, 1996.

Goody, Jack and Ian Watt. "The Consequences of Literacy." *Comparative Studies in Society and History* 5, no. 3 (April 1963): 304–345.

Halverson, John. "Goody and the Implosion of the Literacy Thesis." *Man, New Series* 27, no. 2 (June 1992): 301–317.

Harris, Jessica B. *High on the Hog: A Culinary Journey from Africa to America*. New York: Bloomsbury, 2011.

James Beard Foundation. "The James Beard Foundation's Mission Is to Celebrate, Nurture, and Honor America's Diverse Culinary Heritage through Programs That Educate and Inspire. James Beard Foundation." Accessed September 1, 2014. http://www.jamesbeard.org/about.

Laudan, Rachel. *Cuisine and Empire: Cooking in World History*. Berkeley, CA: University of California Press, 2013.

Lewicki, Tadeusz. *West African Food in the Middle Ages: According to Arabic Sources*. Cambridge: Cambridge University Press, 2008.

Mennell, Stephen. *All Manners of Food: Eating and Taste in England and France from the Middle Ages to the Present*. 2nd ed. Urbana, IL: University of Illinois Press, 1996.

Mignolo, Walter. *The Darker Side of Western Modernity: Global Futures, Decolonial Options*. Durham, NC: Duke University Press, 2011.

Miller, H.D. "The Pleasures of Consumption: The Birth of Medieval Islamic Cuisine." In *Food: The History of Taste*. Edited by Paul H. Freedman, 135–164. Berkeley: University of California Press, 2007.

Montanari, Massimo. *The Culture of Food*. Oxford, UK; Cambridge, MA: Blackwell, 1996.

Neirinck, Edmond, and Jean-Pierre Poulain. *Historia de la cocina y de los cocineros: Técnicas culinarias y prácticas de mesa en Francia, de la Edad Media a nuestros días*. Barcelona: Editorial Zendrera Zariquiey, 2001.

Norton, Marcy. *Sacred Gifts, Profane, Pleasures: A History of Tobacco and Chocolate in the Atlantic World*. Ithaca, NY: Cornell University Press, 2008.

Pilcher, Jeffrey M. *Food in World History*. New York: Routledge, 2006.

Pinkard, Susan. *A Revolution in Taste: The Rise of French Cuisine, 1650–1800* (Cambridge: Cambridge University Press, 2009).

Quijano, Aníbal. "Coloniality and Modernity/Rationality." *Cultural Studies* 21, no. 2 (March 2007): 168–178. https://doi.org/10.1080/09502380601164353.

Ray, Krishnendu. *The Ethnic Restaurateur*. London: Bloomsbury Academic, 2016.

Revel, Jean-François. *Culture and Cuisine: A Journey through the History of Food*. Garden City, NY: Doubleday, 1982.

Rodríguez Monegal, Emir. "Nota sobre Biorges." *Mundo Nuevo*, Abril 1968. https://web.archive.org/web/20071017043300/http://www.archivodeprensa.edu.uy/r_monegal/bibliografia/prensa/artpren/mundo/mundo_22.htm. Accessed June, 2022.

Sahagún, Bernardino de. *Historia general de las cosas de Nueva España.* Edited by Juan Carlos Temprano. Madrid: Dastin historia, 2001.

Trubek, Amy B. *Haute Cuisine: How the French Invented the Culinary Profession.* Philadelphia, PA: University of Pennsylvania Press, 2001.

Waley-Cohen, Joanna. "The Quest for Perfect Balance: Taste and Gastronomy in Imperial China." In *Food: The History of Taste.* Edited by Paul Freedman, 99–132. Berkeley: University of California Press, 2007.

White, Hayden V. *Metahistory: The Historical Imagination in Nineteenth-Century Europe.* Baltimore: Johns Hopkins University Press, 1973.

2
DESENSUALIZING TASTE

The gastronomic writing genre of the nineteenth century formalized the ways of thinking about taste that became possible in Europe as a result of the Enlightenment, the French Revolution and colonial imperial expansion. The conditions of possibility of gastronomic thought were limited by philosophical debates. Desensualization is an unacknowledged consequence of the rationalization of taste for which gastronomy is usually celebrated. Gastronomy's insistence on the establishment of universal rules for the judgment of taste was meant to raise the philosophical standing of the sense of taste, which had been quite low in the dominant strains of Western thought for centuries. However, the establishment of objective rules was only made possible by the suppression of the embodied and affective variables inherent in the experience of taste. Instead of tasting with the senses, gastronomy urged modern subjects to taste with their minds, which was conceptualized as separate and distinct from the body and the senses. The result is a desensualized understanding and experience of taste in which tasting is guided first and foremost by "objective" and "universal" principles that overrule embodied and situated sensory experience. Far from being evidence of the liberation of taste, gastronomic writing provides a record of the process of negotiation and compromise that shaped modern taste. However, the gastronomic compromise was an uneasy one, as evidenced by how humor in gastronomic writing undermines its desensualized rationalist pretensions.

While engaging with philosophers, scholars, medical doctors and cooks, the main perspective of gastronomy came mostly from bourgeois consumers who faced an ever-increasing variety of food commodities coming from overseas colonies and from the growing food industries. Gastronomic texts fashioned bourgeois consumers as new social subjects, called gourmands, who wanted to enjoy the new and abundant food commodities while being seen as rational eaters, as opposed to mere

DOI: 10.4324/9781003331834-3

gluttons. Gourmand status was not dependent on the refinement or on the exploration of the possibilities of the sense of taste. It depended mostly on the acceptance of a discipline defined by a bureaucracy of self-appointed experts. Gourmand subjectivity was achieved through the reading of gastronomic publications that guided consumption with clear rules that were given and received as if they were unquestionable laws of universal knowledge.

Gastronomic texts established attitudes toward food and taste that became normalized among people all over the world who consider themselves to be modern. These attitudes were collectively constructed through books and journals published and read in France, England and other European countries, as well as in the United States. Whereas the first and the largest number of books were published in France, England also made important contributions to the genre. There is evidence that French materials were read in England in French and in translation, and that English authors were read and translated in France. The reader correspondence section of many of the gastronomic journals shows that the texts were read beyond England and France. So, even though France took the lead in the gastronomic writing genre, it was following a logic of thought that resonated with other major imperial powers. The general attitude toward taste was that it needed to be disciplined by reason. A restrained and non-affective concept of taste became the bodily discipline that distinguished bourgeois white European subjects from those that European colonial empires constructed as inferior classes and races. Far from being revolutionary, gastronomic writing refashioned the religious and philosophical call to moderate the enjoyment of taste, although with new rationales, methods and goals.

The gastronomic writing genre demonstrates above all a quest to establish eating as a legitimate field of knowledge. Such legitimacy had been denied in Western thought for centuries. In its search for legitimacy, gastronomers sought to build connections between gastronomy and better established fields of knowledge. That made gastronomic writing notoriously heterogenous. It provides a gourmand's eye view of history, geography, literature, philosophy, science, politics and medicine to identify the ways in which food has proven to be important in those fields. Gastronomic texts were structured as collages of different kinds of discourses imitating the styles of other genres, including historical and geographical accounts, philosophical treatises, literature, dictionaries and encyclopedias. They also include plenty of aphorisms, anecdotes and food trivia. By presenting the interest in food as relevant for learned society, gastronomy gained a modicum of respectability. While there are important differences between gastronomic texts written by different authors depending on their time period, local political perspectives and personal idiosyncrasies, it is possible to consider them collectively to identify defining traits that rise above such differences. The gastronomic writing genre is in general defined on the one hand by its submission to the stringent limitations that rationalism and aesthetics placed on taste and sensing, which produced the strikingly desensualized and non-affective modern notion of taste, and by a self-deprecating humor that ambiguously challenges such limitations, on the other.

The Gastronomic Compromise

Gastronomy constructed modern taste under the constraints established by modern rationalism and aesthetics, and by the ideological needs of capitalism and colonial imperialism. These constraints resulted in a notion of taste that is peculiarly desensualized and affectless, and which has served to delineate and enforce social and racial hierarchies. Gastronomic texts attempted to give the gravitas of natural scientific laws to a series of guidelines that served to discipline taste to follow a code of behavior suitable for bourgeois patterns of social and global order. The notion of taste constructed by gastronomy is in line with the shift in the approach to the control of sensuality in modern societies in which curbing behavior was internalized rather than externally imposed. In Europe the sense of taste was consistently dismissed or condemned in intellectual and religious thought, but in the early modern period knowledge about food and taste began to be valued as a way of controlling appetite. Gastronomic texts document a process of compromise in which a restricted notion of taste was normalized for the sake of establishing gastronomy as an acceptable field of knowledge.

The enjoyment of the pleasures of taste had been hampered in Europe by a long tradition of religious and philosophical thought that looked down on the body and was distrustful of sensory experience. Modern rationalism and aesthetics provided an opportunity for the valorization of the epistemic capability of taste, but they posed new obstacles to gustatory enjoyment. The emergence of gastronomic writing showcases the way in which those interested in validating the pleasures of taste attempted to appease the accumulated baggage of hostility toward this sense in European thought. Nineteenth-century gastronomic texts show the uneasy compromise of gastronomers, in which they both submitted to and challenged the limited space that modern thought had opened for taste in European intellectual culture. Gastronomers imitated and engaged with philosophy and other established fields of knowledge but, as Carolyn Korsmeyer put it, "the interest between philosophy and gastronomy was one way."[1] Gastronomic writers enthusiastically attempted to insert discussions about taste into philosophical discussions as more than a metaphor. However, they only managed to carve a marginal, segregated space for themselves. While gastronomy slavishly imitated other fields of knowledge, the other fields of knowledge continued to discount gastronomy as illegitimate.

Modern rationalism and aesthetics, enduringly defined by René Descartes and Immanuel Kant, established the philosophical parameters that gastronomy had to negotiate. Different strands of modern rationalism agree on the notion that knowledge is established by reason, understood as disembodied, while sensory perception is tied to the body and is therefore conceived as providing only subjective sensations that are next to useless in the production of objective knowledge. This created a longstanding separation between mind and body in the Western philosophical tradition, in which the body and the senses are inferior to the mind. Practices that conform to the rationalist model of knowledge production became highly valued, while those that do not were assigned to the realm of the feminine, lower class, primitive or animal, which were cast as inferior. Anything related to sensory perception, insofar as it is tied to the body, was considered irrelevant in the pursuit of knowledge and truth, and therefore unworthy of intellectual culture. This was a bleak context for the development of a sophisticated taste culture. Taste was considered not only irrelevant but also dangerous because it provides pleasures that could lead to an uncontrollable excess that was harmful to the body and to society.

The discourse of aesthetics emerged in the eighteenth century, in many ways as a response to rationalism and its failure to account for and control sensory experience. As Terry Eagleton has argued, reason had to find a way of penetrating the world of perception without putting at risk its own absolute power.[2] Aesthetics extended the reach of Enlightenment rationality, becoming a prosthesis to reason that attempts to account for what was understood as the inferior logic of the world of perception and sensory experience.[3] Aesthetics validated the senses of sight and hearing as purveyors of the aesthetic pleasure of beauty, which was defined as disinterested, in contrast to the lower bodily pleasures of touch, smell and taste. While taste became a metaphor for aesthetic discrimination, physical gustatory taste continued to be shunned. The acceptance of the notion of aesthetic taste was a corrective to the idea that only reason has evaluative powers. However, it stressed the aspects of visual and auditory perception that make sight and hearing compatible with the concept of disembodied reason – detached perception that is conducive to objectivity – while leaving aside many of their more distinctive aspects. It also downgraded taste, smell and touch even further. The aesthetic requirements of detached perception and disinterested pleasure could hardly be fulfilled by the sense of taste. Taste requires not only contact with the perceived but its actual assimilation into the body of the perceiver. Taste is also rarely disinterested since the act of eating is obviously a need for survival.[4] Gastronomic writers made taste meet these requirements by stressing its objective component and suppressing its sensuous subjectivity.

In the nineteenth century, gastronomic texts were aimed at getting gustatory taste a more dignified position in modern philosophy and culture. Taste sensations had to be submitted to the procedures established by the aesthetic judgment of metaphorical taste, which insisted on universal principles. Conforming to rationalist and aesthetic strictures led to the suppression of the full sensuality and subjectivity of the experience of taste. The concept of taste that emerged in this context was driven by a disembodied mind that was more interested in the establishment of objective knowledge than in bodily pleasure. The subjectivity of taste was sacrificed in favor of establishing as universal the contingent taste preferences of mostly Parisian elites. When taste differences were acknowledged, it was only to be ordered into an exclusionary schema in which those that did not agree with the supposedly universal standards were cast as tasteless. As Denise Gigante put it in the introduction to her anthology of nineteenth-century gastronomic texts, "if one aspired to the 'rational enjoyment' that modern gastronomers were after it was necessary to discipline one's palate according to principles."[5] Since rational deliberation was not considered possible without principles, gastronomy was keen on providing them. Failing to submit to this gastronomic aesthetic discipline would have barred access to modern bourgeois subjectivity and sociability. The experience of eating as a sensuous way of knowing and enjoying was reduced to the task of judging foods as objectively good or bad, according to criteria previously established by self-appointed experts. Rationalism and aesthetics thus imposed an epistemological burden on taste and an ethical burden on gastronomy: taste was to be valued only if it served to establish objective knowledge, and gastronomy had to accept the moral burden of curbing appetite and pleasure, which were considered antisocial and uncivilized. Gastronomic rule rebranded the Christian morality ideals of truth and moderation in the guise of modern self-governance.

Gastronomic writers were the founding bureaucrats of the modern discipline of taste. Their desire to normalize taste meant that they had to disregard the variability of the experience of taste and control its sensual physicality. Gastronomic writers were more attached to the rationalism of the Enlightenment than to the unrestrained passions favored by literary Romantic writers.[6] The position of gastronomers is somewhere between the disavowal of sensory pleasure espoused by the Church and philosophers like Jean-Jacques Rousseau, on the one hand, and the embracing of the more embodied and subjective sensuality favored by philosophers like John Locke, Denis Diderot, Charles Fourier and Jeremy Bentham, on the other. Gastronomers stayed away from philosophical ideas available to them that could have helped them to avoid suppressing the subjectivity and physicality of taste. These ideas included the rejection of the objectivity of taste and the championing of the need to better gratify the senses for a healthier social order. Espousing these ideas would have resulted in a richer notion of taste, but it would also have put into question the possibility or desirability of a discipline of taste. There would have been no need for gastronomers as bureaucrats of taste. Gastronomy was driven by a will to control the taste experience by establishing inflexible universal rules, even if that meant neglecting the many dimensions of taste that exceed the demands of universality.

While the approach of gastronomers to taste was defined by moderation and control, other thinkers had more complex and flexible approaches to sensory experience. For example, Locke in 1690 explored the complexities of taste and the conditions of human knowledge in his *Essay Concerning Human Understanding*. In a brief discussion of the taste of pineapple, a fruit that he likely never tasted, Locke implied that taste is defined by more elements than the materiality of food. As Sean R. Silver put it in his analysis of the *Essay*, travelers' accounts had given people in England a taste for the pineapple before they could have a taste of it.[7] Silver explains that Locke sure enough made the point that only people who have eaten pineapple know how it tastes, but he also acknowledges that taste is bound to the cultural ideas and expectations that surround it.[8] Even though the pineapple, like all food objects, has specific sensory qualities that can only be known through the act of eating, there is no objective taste unmediated by cultural ideas. People who had never tasted a pineapple, but had heard or read about it, desired it and pronounced it delicious. This understanding of taste is an example of the complexity that the discourse of gastronomy disregarded with its insistence on objectivity and universality. Rather than advocating in favor of a fuller understanding of the experience of taste that includes both the material and the cultural, and the subjective and the objective, gastronomers opted to excise from their notion of taste any aspects that were seen as incompatible with the rationalist disembodied concept of reason.

Over the course of the Enlightenment, many other writers defended the pleasures of taste without negating its subjectivity and physicality. The preface to the 1750 cookbook *Les Dons de Comus* praises the art of cooking and argues that cuisine should not be blamed for the ill effects of intemperance and of ignorant cooks.[9] The author of the preface confidently praises the work of knowledgeable cooks who are well versed in chemistry. The preface refers to cuisine as "a very free art that does not have any other rules other than taste where ideas and impressions are infinitely varied."[10] The author of the preface is keenly aware that everybody has a different taste and that no cooking method can satisfy everybody.[11] This author did not see a conflict between accepting the variability of taste and taking advantage of science to fashion new ways of cooking. Another author who did not think that embracing sensuality negated his claims to reason was Diderot, chief editor of the *Encyclopédie* (1751–1772). Diderot was a self-confessed sensualist who defended the growth of appetites and pleasures as a sign of social progress.[12]

Even after *The Physiology of Taste* (1825) by Jean Anthelme Brillat-Savarin had become established as the authority on the "science of taste," non-gastronomic writers continued to advocate for a fuller understanding and satisfaction of taste than the one espoused in the gastronomic genre. Charles Fourier, Brillat-Savarin's relative and contemporary, wrote what amounts to a defense of the right to individual taste preferences, which he thought was being trampled by gastronomy. He contrasted how an old hen as a food object creates discord in his contemporary civilization, which he despised, but fosters bonds in Harmony, his ideal future stage of humanity. In civilization, people of means who would like to eat old hens are discouraged, if not disparaged, because such food has been ruled undesirable. On the other side of the social spectrum, the poor have no option but to eat old hens even if they do not like them. In civilization, old hens cannot be enjoyed by those who have a taste for them; it is only a food to be endured by those with fewer options. In contrast, in Harmony the people who prefer old hens to tender poultry would not be teased about their "bizarre taste," and they would get together for group dinners featuring their preferred food. These group dinners would unite and raise the prestige of producers, preparers and consumers of old birds.[13] In Fourier's view, gastronomy deprives both rich and poor of gustatory pleasure because food choices are dictated by general rules and class status rather than by personal taste preferences. Bentham is another philosopher contemporary to Brillat-Savarin who made a more radical defense of sensual pleasures than gastronomic writers. Bentham criticized all ideologies that placed limitations on the liberty of taste, particularly when it comes to food and sex.[14] He did not distinguish between higher and lower pleasures and did not consider any particular taste preferences to be superior to others.[15] From Bentham's and Fourier's perspectives, gastronomy looks like a tyranny over gustatory pleasure.

Fourier was interested in managing the passions or appetites, but his approach was one of control by fulfillment rather than by deprivation. He argued that problems are caused not by excessive pleasure but by how rare pleasure is in civilization.[16] He saw his contemporaries as "a public moral and uniform in its tastes, eating only to moderate their passions, and forbidding themselves all sensual refinement, for the benefit of repressive morality."[17] He directly attacked his famous relative: "Savarin was a simplistic ignorant of gastrosophy … the art of combining the refinements of consumption and preparation … the art of binding all the branches of the system of production and subsistence."[18] Fourier's "gastrosophic" ideas were more comprehensive and less classist than Brillat-Savarin's gastronomy. Fourier claimed that whereas in civilization love of eating leads to laziness and excess because consumers have not labored to produce the food, in his ideal society everybody would participate in food production and have access to choice commodities.[19] Epicurism, understood as having a high degree of gustatory refinement and the ability of gratifying it, would be everyone's right. When describing his plans for the education of children, Fourier argued against forbidding them the dainties that they tend to prefer.[20] His education plan integrated agriculture, gastrosophy and cooking as a way of developing the senses of taste and smell, in a complete program that would prevent the excesses of civilization.[21] Fourier's criticism of gastronomy clearly defined it as just a joyless and unrefined way of controlling pleasure.

Fourier's gastrosophic education ideals contrast sharply with the more influential ones that were espoused by Rousseau in 1762. In *Émile*, Rousseau prescribed for children a diet of moderate amounts of fruit, dairy and bread, in the interest of avoiding arousing their sensuality.[22] Rousseau was one of many Enlightenment philosophers who were hostile to elaborate cuisine arguing that the sensuality of exciting dishes corrupted the physical, moral and social body. Because of the need to appease the continued objections to the cultivation of taste set forth by philosophers like Rousseau and his followers, moderation became a core value of gastronomy. Nineteenth-century gastronomers like Brillat-Savarin developed their position as a middle ground between Rousseau's ascetic moralism and Fourier's socialist sensualism. Gastronomers distanced themselves from Rousseau by proposing that the gratification of the sense of taste was a morally and intellectually acceptable endeavor, but they did not go as far as embracing Fourier's call for the exploration of the subjectivity of tasting pleasure unencumbered by universal rules. Even in their own geohistorical context, gastronomic writers were not the champions of sensory pleasure that they are reputed to be.

Gastronomers established interest in food as intellectually acceptable but only by approaching sensuality as something that needed to be tamed and justified by moderation, reason and science. The philosophical split between minds and bodies, and the requirements of objectivity, universality and disinterestedness, made it impossible for gastronomers to celebrate and explore the many aspects of taste that go beyond any notion of objectivity. Gastronomers compiled in their volumes all kinds of food-related information and presented it as a corpus of objective knowledge that should guide eating behavior. To establish the legitimacy of gastronomy as a new field of knowledge, according to how knowledge was defined in modern Europe, gastronomy could not take shape as a practice seeking to refine and enhance gustatory pleasure. The literal meaning of the word "gastronomy" is "rules of the stomach." This word does not evoke sensuality and pleasure. Rules are about curbing rather than enhancing pleasure, while the stomach is about the physiological process of digestion rather than the sensorially pleasurable aspects of ingestion. From an outsider's perspective, gastronomy can look like a crass taste culture, with its emphasis on the stomach as the site of digestion, rather than on the refinement of the senses. It can also seem oddly depersonalized and desensualized. The reduction of the experience of taste to an operation of a disembodied mind and a machine-like stomach that follows bureaucratic standards required the subordination of the freedom to explore the individual variations of sensual experience and pleasure. To paraphrase Foucault's famous point regarding sex, the modern West developed a *scientia gastronomica*, rather than an *ars sensualis*.[23] Gastronomy is the result of a compromise between the desire to validate the pleasures of taste and the need to appease the longstanding philosophical hostility against such pleasures in Western thought.

Scientia Gastronomica, Not Ars Sensualis

The demands of rationalism and aesthetics produced gastronomers and gourmands as subjects with a specific relationship to their bodies and their senses. As the literal meaning of term "gastronomy" indicates, the gastronomic body is defined by its well-ruled stomach, not by the refinement of its sensory organs. Since the subjectivity of the senses was seen as a liability that could compromise the intellectual respectability of gastronomy, gastronomers were committed to the construction of a sense of taste that was objective. They grounded the specificity of their practice on digestion and on taste understood as a strictly physiological and chemical process. Digestion was by far more amenable to the idea of general rules than the experience of taste could ever be. Digestion helped to relate gustatory concerns to the rationalism of the modern sciences, while rules turned tasting into a task akin to the appreciation of art as defined by the philosophy of aesthetics. A significant portion of pages in gastronomic texts is devoted to the establishment of rules for cooking, hosting and eating, and to their justification as aiding, or at least not disrupting, proper digestion. Digestion was a main scientific and public concern in the eighteenth century.[24] Focus on digestion is what saved gustatory concerns from illegitimacy in modern culture since it made cooking and eating knowledge important both socially and scientifically. This knowledge was socially relevant because it curbed excess sensuality that would harm both digestive and social health, and it had scientific relevance as a supplement to medical knowledge.

Cookbook authors from the seventeenth century onward had already begun the kind of writing about food that came to characterize the gastronomic genre in the nineteenth century. The preface of many cookbooks included historical accounts, discussion of scientific and medical principles relevant for cooking, aphorisms, food-themed literature and anecdotes. They often took the format of an academic treatise. Cookbook preface authors were also the first ones to develop the defense of cooking as an art and science, and they took steps to shape their cooking practice to conform to the shifting dominant perspectives on science and aesthetics. Gastronomic writers were compilers of information from many different sources. They shaped their material in an authoritative style and tone that helped it become established as enduring depositories of food-related knowledge. Gastronomic writing showcases the victory of a bourgeois approach to taste over the aristocratic one. While aristocrats devoted a lifetime to the development of a discerning taste, the bourgeois only had to follow gastronomic guidelines. Gastronomic writers built on the strategies developed by cooks to systematize cooking and used them to systematize their approach to the experience of taste.

In 1801 Joseph Berchoux published the long poem *La Gastronomie*, which is considered to be the first appearance of the term "gastronomy" in print. The poem introduces gastronomy as a new discipline in search of disciples. The voice of gastronomy summons eaters to attend its school:

> You who up to this day, stranger to my laws,
> Have followed your tastes without method and without choice;
> Who, in your appetite ruled by habit,
> Do not know the art that I study,
> My voice is going to dictate you important lessons:
> Come to my school, oh my dear infants![25]

The condescending and authoritarian voice of gastronomy addresses its prospective students as ignorant children. Gastronomy is presented as the sole possessor of a new but indispensable knowledge. Readers are treated as delinquent regarding laws they did not even know existed and in need of gastronomy to discipline them. In the same vein, in 1805 Charles-Louis Cadet de Gassicourt published the *Cours gastronomique*, a book in which readers get lessons about food history and geography through a series of conversations between different characters, including one addressed as "Professor." The course stresses the importance of knowing enough chemistry to avoid eating in ways that could lead to bad digestion.[26] Finally, the course outlines a canon of food-related books to furnish a gourmand library. Berchoux and Gassicourt wanted their readers to become aware of their need for the guidance that gastronomy offered. Eating was no longer to be considered something that could be done without guidance from the newly minted class of bureaucrats of taste. Eaters had to submit their senses and pleasure to the purportedly disembodied rules of gastronomers.

One of the most influential gastronomic publications was Alexandre-Balthazar-Laurent Grimod de la Reynière's *Almanach des Gourmands* (1803–1812). Grimod was an aristocrat who took on the task of begrudgingly passing down the standards of aristocratic taste to the affluent members of the bourgeoisie, whom he saw as having money but no class. In his series of publications, Grimod adapted the standards of taste to accommodate the innovations in food products that were the result of capitalist and colonial production and commerce. Often credited with inventing food criticism, Grimod systematically presented himself as an authoritative mediator between food producers and consumers. He convinced producers to send him samples for his approval and told readers which of the products were worth buying, making both sides dependent on his judgment. Rather than exploring their own preferences, Grimod's readers were eager to accept his judgment as objective expertise.

While the enthusiastic reception and proliferation of gastronomic publications suggest the existence of a readership receptive to gastronomic teachings, gastronomy continued to face resistance. In 1806, Berchoux's poem was refuted by Jean-Baptiste Gouriet's poem *L'Antigastronomie*. This poem aimed to steer people away from gastronomy by mobilizing ideas from the long history of philosophical, moral and medical calls for eating temperance. Paraphrasing Rousseau's *Émile*, the poem includes a critique of the exploitative global relations of power that are necessary to provide the gastronomic tables where "a single man devours towns and nations."[27] Gastronomic writers, however, were unmoved by the atrocities committed by European colonial imperialism to provide the bounty for their tables. Gourmands felt entitled to the alimentary wealth of the planet on account of the superiority that modern/colonial race discourse and power relations afforded them.

In 1817, the doctor William Kitchiner published *The Cook's Oracle*, a book that saw many editions and became a classic in England and the United States. This cookbook was intended to provide recipes for gourmand-pleasing foods that would not harm the body or the budget. If Grimod's authority came from being an aristocrat, Kitchiner's authority came from his knowledge of medicine. Indulging in the foods recommended by Kitchiner would please the palate without harming digestion and therefore losing the claim of being a rational eater. Kitchiner's recipes were cooked and tested by a "*Committee of Taste* composed of some of the most illustrious gastropholists of this luxurious metropolis."[28] The readers of this book were expected to trust the medical judgment of the author and the approval of a group of experts trained in the London fine dining scene.

When Brillat-Savarin published his influential book in 1825, gastronomers no longer needed the approval of either aristocrats or doctors, because they had already internalized aristocratic and medical guidelines as their own. Bourgeois gastronomers now had a confident hold on the aristocratic and medical standards of taste and saw themselves as bearers of a gastronomic knowledge more authoritative than any other. All gastronomic writers considered that they were providing unequivocal truths. More than any other gastronomic writer, Brillat-Savarin was convinced that his work was the definitive source on the subject of food and eating. On the preface to his book, he revealed his ambition to be recognized as a man of science and his fear that he would be dismissed as just a compiler.[29] His strident confidence was tempered by an awareness that his bid to turn gastronomy into a science was not entirely convincing.

In 1835, Thomas Walker published *Aristology*, a book that outlined the principles that he thought should govern all dinners. Walker is often dismissed as just another British follower of French models. The *Aristology* is similar to Brillat-Savarin's *Physiology*, in that they both were proposed as comprehensive depositories of rules and in that their authors had no other credential than being gourmand bourgeois consumers. But the rules that Walker advocated were specifically geared toward a no-nonsense or simple enjoyment of excellent food without allowing rules of politeness to get in the way. He associated excessive politeness with the vulgar rich, which he did not want to imitate.[30] Walker is an example of the confidence that the bourgeois were increasingly feeling as a hegemonic class capable of establishing its own rules. His authority to establish standards of taste came from his sole focus on having an eating experience unburdened by politeness. But all the same, he presented his views in the form of rules to be followed. Gastronomers might have disagreed on specific rules, but they all thought that eating know-how could only take the form of laws.

Brillat-Savarin defined gastronomy as "the intelligent knowledge of whatever concerns man's nourishment."[31] This definition reaffirmed the idea that there are intelligent and non-intelligent ways of eating and that gastronomic knowledge was needed to not be considered an ignorant and undisciplined eater. The main purpose of gastronomic writing was to establish the gourmand as something different from the glutton. Grimod had granted that both the glutton and the gourmand have a big appetite but argued that the senses of the gourmand are superior because they are more delicate.[32] Brillat-Savarin, on his part, defined gourmandism as "an impassioned, considered, and habitual preference for whatever pleases the taste" and insisted that gourmandism is the enemy of overindulgence.[33] Both the glutton and the gourmand have a big stomach but, unlike the glutton, the gourmand does not eat indiscriminately. He uses his superior senses and knowledge to judge foods and consume only the best. While all gastronomic authors repeatedly paid lip service to the idea of temperance, their texts provide ample evidence that they were not hostile to the idea of excess as long as the excess food was considered excellent. Since excess quantity was justified by high quality, the gourmand was little more than an affluent glutton.

Brillat-Savarin, by his self-description, was overweight.[34] He saw the overindulgence that gave him a big belly as a characteristic of civilized peoples: "As for us, citizens of the old and new worlds who believe ourselves to be the finest flower of civilization, it is plain that we eat too much."[35] Brillat-Savarin dedicates separate chapters to obesity and its treatment, and he ambitioned writing a whole book on the topic. While he granted that obesity was a problem, his approach to controlling it had more to do with fad diets than with the temperance that was supposed to define the gourmand. Rather than eating moderately, Brillat-Savarin recommended taking it easy with grains and starches, drinking 30 bottles of Seltzer water every day during the summer, and wearing an anti-fat belt.[36] He and other gastronomic writers would have been pleased by the invention of a pill that would allow them to eat as much as possible without feeling full or gaining weight. Moderation was a sad duty half-heartedly accepted by gastronomic writers in order to make interest in food acceptable for polite society, but they were gluttons at heart. The gastronomic emphasis on digestion gave gastronomic writing a veneer of scientificity. It was driven by a glutton's desire to free eating from the constraints imposed by the body, which limits enjoyment by becoming full, fat and sick. The gourmand's body is not defined by refined senses but by a stomach always threatened by indigestion.

The gastronomic discipline was based on a mechanical notion of the body and the taste experience. Gastronomers saw their bodies as machines that tasted and digested efficiently. For Brillat-Savarin, the superiority of "man" over other animal species was visible in his powers of ingestion and digestion:

> As soon as an edible body has been put into the mouth, it is seized upon, gases, moisture, and all, without possibility of retreat.
>
> Lips stop whatever might try to escape; the teeth bite and break it; the tongue mashes and churns it; a breathlike sucking pushes it towards the gullet; the tongue lifts up to make it slide and slip; the sense of smell appreciates it as it passes the nasal channel, and it is pulled down into the stomach to be submitted to sundry baser transformations without, in this whole metamorphosis, a single atom or drop or particle having been missed by the powers of appreciation by the taste sense.[37]

Brillat-Savarin's account of the process of eating and tasting considers taste and smell appreciation as a strictly physiological aspect of the machine-like efficiency of the mastication and digestion processes.

Brillat-Savarin's perspective on the eating process is heir to the preoccupation with digestion that characterized Enlightenment culture. E.C. Spary argues that philosophers gave a lot of importance to digestion, which was seen as the main way in which external matter determined the characteristics of the thinking self.[38] Digestion was thought to shape both body and mind, and as such it was important to understand it. In 1739, Jacques de Vaucanson exhibited a defecating duck automaton. The duck was a mechanical device that would ostensibly eat and defecate, but it was later exposed as a deception. The idea behind the duck was to demonstrate that digestion was a chemical rather than a mechanical process.[39] Many other attempts to model digestion experimentally were made up to the early nineteenth century. This interest among the scientific community and the general public demonstrates the importance that digestion had in a context where diet was believed to determine moral character and racial identity and in which food habits were changing fast due to the increasing availability of foods from overseas colonies. People were anxious about how consuming foods from foreign landscapes would affect their health and sense of self. It also shows why association with digestion rather than with eating pleasure was gastronomy's best bet in its search for legitimacy in the modern thought context.

The defecating mechanical duck was only one of many automata that became popular throughout the Enlightenment. The human body was understood to work as a machine, and building machines was a way of figuring out how the body works.[40] Cartesian philosophy supported a mechanistic view of the physical universe, which included human behavior.[41] But, as a dualist, Descartes was also committed to the independence of mind and to free will. Ideally, human behavior would be determined by rules as natural laws. But since, unlike machines, people could choose whether to follow the rules of behavior or not, they deserved moral praise or condemnation according to their choice.[42] The automatization of adherence to rules of behavior relates to what Norbert Elias called the civilizing process, in which people internalized a discipline and followed it without noticing because they felt it as natural.[43] Gastronomy was shaped by this ideal of socialization and civilization conceived as rational automatization. Cooks and gastronomic writers formalized the processes of cooking and eating so they could unfold in a predictable, controlled way that was eventually considered natural and rational. The experience of eating following gastronomic rules was a show of "proper" and "civilized" behavior, but it negated the validity and curbed the intensity of personal sensory experience. Descartes and others wrote fencing manuals, which transformed the angry impulses of a sword fight into a choreography disciplined by mathematical formulas. Gentlemanly fencers were automata going through the motions, mimicking violence while emptying its affective core.[44] Likewise, gastronomy turned eating into a formalized social event in which eating is to be enjoyed in a controlled, rational, automatized way. Gastronomy placed polite behavior and disciplined bodies ahead of gustatory pleasure. The gastronomic body, conceived as a digesting machine, was not endowed with subjective senses capable of affecting and being affected by taste as an experience.

From the perspective of those interested in the refinement and expansion of the pleasures of taste, the success of gastronomy was pyrrhic at best. Gastronomy did very little for the validation of taste as an affective embodied experience. Gastronomy is to taste what sexual education is to eroticism. The focus of sexual education is responsible reproduction and the avoidance of sexually transmitted diseases, not erotic pleasure. The focus of gastronomy was digestion, not tasting pleasure. Both sexual education and gastronomy are disciplining knowledges aimed at controlling pleasure. Gastronomy is a discipline in the sense of disciplining and surveilling behavior. In its zeal to gain scientific and aesthetic credentials, gastronomy muffled the very pleasure it was supposed to be championing. Gastronomic writing, together with diet and nutrition science, disciplined taste in a manner similar to how, according to Foucault, sex was disciplined by placing it into discourse.[45] Placing sex or taste into discourse is neither repressing nor liberating. It is part of the process of creating ways of talking about and participating in sex and taste experiences that are considered appropriate for the functioning of the modern social order. Readers of gastronomy were expected to model their dining room behavior and their gustatory preferences after the models provided in the texts. The price paid for membership in the polite class was a taste experience determined by the rules of a taste bureaucracy.

The relationship between sex and eating as socially dangerous drives was very much in the minds of gastronomers. In gastronomic writing, sex was often invoked as a darker pleasure that made the pleasure of taste acceptable as the lesser of two evils. This theme was already present in Berchoux's *La Gastronomie*. In this poem, the voice of gastronomy addresses young men and invites them to steer away from women who will betray them and to instead trust the pleasure of gastronomy, which will never let them down.[46] The theme continues in the first volume of Grimod's *Almanach*, where a "true gourmand" compares at length the pleasures of good food with the favors of a woman. He concluded that the pleasures of good food are undoubtedly superior.[47] In the sixth volume of the *Almanach*, the author makes clear that women are welcome at the daily table, but they should be excluded from "erudite dinners."[48] Brillat-Savarin considered that women are gourmandes but in an instinctive rather than rational way, because it is favorable to their beauty.[49] In general, the true gourmand is male, often celibate, and frequently misogynistic.

In 1804, shortly after the publication of the first volume of Grimod's *Almanach*, a vaudeville play poking fun at the gourmand's preference of food over sex was written and staged. It has been suggested that the main character was indeed a spoof of Grimod himself.[50] In the play, called *L'École des Gourmands*, a gourmand is tricked into giving up his pretensions to marry his young pupil. The trick consisted of making the gourmand believe that marrying his pupil would make him lose the services of his talented cook. The gourmand realized that a wife would only be an impediment to gourmandism, and he cheers himself up from the loss of the prospective bride by exclaiming that life is a good dinner in which getting married is only an appetizer.[51] The existence of the play suggests that the figure of the gourmand was already well known and recognizable by theater audiences. It is notable that what makes the gourmand a recognizable figure was not the learned status that they aspired to but their simple-minded privileging of food above all else. In Brillat-Savarin's book, which represents a consolidation of ideas that had been circulating for decades, physical desire was classified as a sixth sense, but it was characterized as a sense less cautious and prudent than taste.[52] Gastronomy was constructed as a male endeavor that was considered less socially pernicious than sex. This gave gastronomy the moral justification that was needed.

Humor and the Discomfort with the Gastronomic Compromise

The gastronomic genre mimicked the contents and styles of established fields of knowledge. In their search for legitimacy, gastronomic writers closely imitated the conventions of other disciplines. They wrote the self-mythologizing history of gastronomy as characteristic of the end point of the linear march of human progress. They gave rules to gastronomy in imitation of the way in which all arts were supposed to have rules. Cooks, on their part, imitated art and architecture by privileging the visual aspect of food. Another field imitated by gastronomy was literature. No gastronomic book or magazine was complete without a literary section with food- and stomach-themed poems and plays. Gastronomic literature tended toward the humorous and satirical, honoring the long tradition of food needing comic support to appear in literary texts. This underlines the conservatism of gastronomic writing when compared to instances of Romantic literature in which food was often used to explore the limits of human experience, including pleasure. But philosophy and science were clearly the discourses whose approval gastronomers craved the most, given that they had been responsible for the low status of taste and cooking in Western culture.

Gastronomic texts use scientific and philosophical vocabulary and formats just as a façade. For example, Grimod's essay "On Mustard and on Syrups, Philosophically Considered" uses the word "philosophy" in the title to great effect, but the contents are limited to recommending different producers of these items after cursorily mentioning some information regarding their history and medical properties.[53] Similarly, the title of Brillat-Savarin's book cries out loud for scholarly recognition: *The Physiology of Taste or Meditations on Transcendental Gastronomy*. Each word is a cue for the public to see this book as a serious philosophical and scientific treatise.

Brillat-Savarin's book is composed of 29 essays, which he called "meditations" and in which a specific subject of gastronomic interest is discussed in a style reminiscent of philosophic and scientific writing. The contents combine established and often obvious knowledge with idiosyncratic and at times insightful perspectives, opinions and anecdotes, all packaged with the formal tone of a professorial lecture. Taste itself is the subject of only roughly 12 pages, filled with mostly obvious observations like this definition of taste: "in physical man it is the operation by which he distinguishes various flavors."[54] As a keen observer, he realized that taste and smell work together and that the sensation of taste develops in a sequence from an initial to a final impression and culminates with reflection. But we can safely assume that many other keen observers had come to the same realization throughout human history. Gastronomers attempted to pass their pedestrian essays as philosophical and scientific writing. What made them scientific and philosophical in their minds was the fact that they were written with an air of authority as if they contained newly discovered universally valid knowledge. The results bring to mind Julio Cortázar's short story "Instructions on How to Climb a Staircase" (1962), in which the explanation to do something that everybody knows how to do has an uncanny and humorous defamiliarization effect.[55]

Gastronomic writing highlighted the importance of food as it relates to all existing fields of knowledge. The imitation of the conventions of the different fields of knowledge was a way in which gastronomers attempted to validate their interest in food as more than gluttony. However, their imitation of the structure and language of other fields often had an – intended or not – humorous effect. Many scholars of gastronomy have chosen not to mention the humorous aspect of nineteenth-century gastronomic texts, but others have at least acknowledged it. Roland Barthes, for one, said that Brillat-Savarin's was an "irony of a science."[56] I argue that the irony and humor of gastronomic texts is constitutive rather than marginal. Gastronomic writing, for all its authoritative pretensions, was really a site of struggle and compromise. Gastronomers knew what it took to turn food into a legitimate topic for lettered culture in their context. Their writing was key to the establishment of a modern notion of taste that did not transgress the parameters of acceptability established by rationalism and aesthetics. However, while they attempted to comply with the philosophical prescriptions, they seemed aware that the parameters did not allow them adequate space to account for the experience of taste in all its complexity. The decidedly embodied and affective aspects of taste that gastronomy was forced to renounce as an undesirable excess kept coming to the surface. Some of the writers embraced this ambivalence, while others tried but failed to mask it. Humor was a strategy used by many gastronomic writers to say the kinds of things they knew it would not be acceptable to say while aspiring to establish the difference between gluttons and gourmands.

In the introduction to the first volume of the *Almanach*, Grimod explains the context in which he is writing. He writes to educate the new rich who, lacking refinement, have turned toward purely animal pleasures.[57] Grimod begins a reflection about how the pleasures resulting from cuisine have always had an air of distinction in society and that the stomach has a surprising influence over moral destiny.[58] However, he stops this reflection to avoid getting lost "in metaphysical discussion, which may not be understood by all those for whom we write, and that belong to philosophy more than to cuisine."[59] He quickly moves on to the focus of his book, which is "to guide and enlighten gourmands in the labyrinth of their aperitive pleasures."[60] It is clear that Grimod, as a member of the aristocracy, considers that he knows how to enjoy the pleasures of the table in a refined way. Unlike the myth perpetuated by the narrative of gastronomic progress, gastronomy was not the first time that people in France or elsewhere approached eating with a heightened intellectual and aesthetic sense. It is also clear that he thinks that his audience lacks this knowledge. He is grudgingly passing the torch of the pleasures of refined eating to the new powerful class. His disdain for the bourgeois can be seen in his use of irony, but it can easily be missed. The frontispiece of the first volume of the *Almanach* is an illustration with a caption that explains that it depicts the library of a gourmand, in which all the books have been replaced by choice food and wine supplies. The illustration is at the same time a provocative assertion of the validity of a food epistemology and a critique of the new affluent gluttons that are about to be enlightened by his book.

Many of Grimod's readers did not notice the satirical stance of the *Almanach*. In the sixth volume, Grimod shares with readers his reaction when the most commented and reprinted part of the previous volume was the recipe for an "unparalleled roast."[61] The unparalleled roast consisted in 17 different kinds of birds stuffed into each other and cooked over 24 hours.[62] Grimod says that he would have been astonished by the success of that section if he didn't already know from experience that the best articles of the *Almanach* are never the ones that receive the most attention. He dismisses his article about the unparalleled roast as silly, peppered with satire and impractical.[63] A reader was so taken by the recipe that he sent Grimod other similar recipes, including the one for "monstrous eggs." The monstrous eggs were made by using pork bladders to shape numerous egg whites and yolks into the shape of a single gigantic egg. While Grimod continues to look down on his readers, he caters to their unrefined taste. Only the occasional use of irony reveals his distance from some of the materials that he published. Whether the unrefined bourgeois noticed the ironic distance or not, it can be seen that at least a few of the readers of gastronomic texts unabashedly embraced an aesthetics of excess that was contrary to gastronomy's stated moderation goals and were not bothered by the disdain with which Grimod addressed them.

Gastronomic texts often include sections that humorously problematize the relationship between gourmands and knowledge. In an essay called "Des langues mortes," Grimod explains that dead tongues are important for the gourmand and for any cultured person, but in the fourth paragraph it becomes clear that he is not talking about languages but about tongues, as in beef and pork tongues.[64] He argues that dead tongues are more important than living languages because they don't lie and hurt. The same book that included this essay contained a brief anonymous anecdote in which a gourmand is having dinner in the company of scholars who were talking loudly and not eating. The gourmand asks them to stop talking so loudly because after a while one does not know what one is eating.[65] These two selections from the book *Le Gastronome français*, published in 1828, use a droll humor to express their resentment toward the privileging of speaking over eating, and knowledge over pleasure. This is even more clearly presented in an anonymous dialogue between the brain and the stomach that was published in the gastronomic journal *Le Gastronome* in 1830. In this dialogue, the stomach confronts the brain, arguing that he is as important as the brain and that the brain depends on him.[66] The closing message of the stomach is that the brain cannot think or understand anything if the stomach does not get food to digest. Gastronomic texts never really stopped showcasing a resentment against a view of knowledge that excludes corporality and sensing, and humor was a safe way to express this resentment.

Gastronomic writing in general kept a ludic ambivalence between renouncing, disguising and embracing gluttony, understood as having an exclusive focus on plentiful and excellent food. Launcelot Sturgeon, one of several British writers who penned gastronomic texts after Grimod but before Brillat-Savarin, is no longer laughing at the bourgeois gourmands but is laughing with them. In the cover of his 1822 book of moral, philosophical and "stomachichal" essays, he identifies himself as "Fellow of the Beef-Steak Club and an Honorary Member of several Foreign Pic Nics."[67] He uses plenty of material verbatim from Grimod, but his attitude is totally different. In the introduction to the book, Sturgeon uses several lines from Grimod's introduction to the first volume of the *Almanach*, but he skips the insulting remarks regarding the new rich. Sturgeon's book is humorous in a more open way, as he wants readers to embrace the reality that stomachs demand attention. In a section on "dinatory tactics," rather than instruct readers in the art of carving at table and give advice regarding polite behavior at the table like Grimod had done in *Manuel des Amphitryons* (1808),[68] Sturgeon advises against allowing etiquette to interfere with the satisfaction of appetite. For example, he recommends avoiding sitting next to the mistress of the house "unless you choose to incur the risk of being forced to waste your most precious moments in carving for others instead of yourself."[69] If put in the situation of having to carve, Sturgeon warns "let not a mistaken notion of politeness induce you to part with all the choice bits before you help yourself."[70] Sturgeon and others defended the right of gourmands to assert their gastronomic interests without hiding them with a veil of politeness.

Brillat-Savarin's book was conceived as a more complete and permanent gastronomic reference than Grimod's almanachs and other gastronomic publications. One of the two stated purposes of the book was "to set forth the basic theories of gastronomy, so that it could assume the rank among the sciences which is incontestably its own."[71] Brillat-Savarin wanted his book to be the definitive guide to the universal rules of gastronomy as a science, and it has indeed been taken as such by many. The tension between the authoritative and the humorous is present throughout the book. The first part of the book, composed of a series of "meditations," are essays written in a scholarly style in which Brillat-Savarin tried his best to sound like a professor. The reason why the meditations seem funny at times is because of the incongruency between the subject matter and the style, which is borrowed from other fields of knowledge. Brillat-Savarin, and gastronomy in general, failed to develop a style and a language appropriate to account for the complexity of the experience of taste. Borrowing from other fields, and particularly from fields that had actively excluded food and tasting concerns, left gastronomy with an ill-fitting template. Brillat-Savarin's book clearly exceeds the template that he had imposed on his material and interests. The second part of the book, almost a quarter of the total number of pages, is called "Varieties," and according to Brillat-Savarin it is composed of anecdotes that would have broken up the main line of thought if scattered through the theoretical part of the book.[72] The introduction to the "varieties" section reveals both a desire to write about his gastronomic experience in a more personal and subjective way and a hesitation and fear of being mocked for doing so. He hoped to be safe "well protected under my philosopher's hood."[73] Brillat-Savarin's academic formalism is only a cover. He wrote following the rules under which it was possible to write a book about food in nineteenth-century Europe. Still, he knew that such rules did not allow him to fully engage with the subject of taste.

Brillat-Savarin's "Varieties" are an attempt to write about food and taste in a more personal, sensual and affective way. This kind of writing, however, is the rare exception and not the rule in gastronomic writing. Like Sturgeon and others, in the varieties Brillat-Savarin shows himself as greedy about good food and intensely affected by the pleasure received by specific dinners. In "Eggs in Meat Juice," Brillat-Savarin relishes the memory of the time when he feasted on eggs scrambled in the juice that he made the cook of an inn steal from the lamb roast of other guests.[74] A great part of the deliciousness of these eggs came from laughing at the unsuspecting guests who would be eating a dry roast. In this anecdote, Brillat-Savarin is not apologizing for his greediness or for unfairly treating others. Adventures like this are what qualify him as a real gourmand, whose misdeed was only possible because he had enough gastronomic expertise to know that all the flavor of meat is in its juices. The difference between a glutton and a gourmand is not that the gourmand eats moderately but that the gourmand has food knowledge. That knowledge, however, could make him as food obsessed and greedy as any ordinary glutton. In this anecdote, the notion of taste at play is more complex than the "scientific" one officially espoused by gastronomy and by Brillat-Savarin himself in the first part of his book. The taste of the eggs was not only an objective property of the eggs and the meat juice, but in this one instance they also had the taste of a mischievous victory. Locke and other non-gastronomic thinkers would have understood it, but this notion of taste as the result of a specific experience did not fit the straitjacket of the gastronomic discipline.

Another one of Brillat-Savarin's varieties shows how different nineteenth-century gastronomic writing could have been if not hampered by the idea of gastronomy as an objective science. "The Curé's Omelet" is about his admiration of an omelet that he heard about from somebody who witnessed somebody else eat it.[75] Brillat-Savarin went to a dinner at the house of the woman who kept the Curé company while he ate what seemed like an exceptional omelet. The description of the omelet given by the woman was very sensuous, even though she did not get to taste it.

> [...] he attacked the omelet, which was round and big-bellied and cooked to the exact point of perfection.
>
> At the first touch of the spoon, the paunch let flow from the cut in it a thick juice which was as tempting to look at as to smell; the platter seemed aflood with it, and our dear Juliette admitted to herself that it made her own mouth water.[76]

During Brillat-Savarin's dinner at the woman's house, the only topic of conversation was the Curé's omelet, and each one of the guests "contributed to make a kind of sensuous equation in the discussion."[77] The anecdote at one level is a criticism of the affluence and hypocrisy of the clergy who enjoy the best foods even while fasting and preaching moderation. But what makes this anecdote remarkable is the sensuous description of the food and the fact that the omelet could be so thoroughly enjoyed by people who did not even see it. The anecdote is followed by instructions to prepare a tuna omelet, which helps readers appreciate how special the Curé's omelet was. It might even inspire readers to reproduce the omelet in their kitchens. This anecdote stands out because descriptions of the organoleptic qualities of food and of their varied physical and emotional effects on the eaters are remarkably infrequent in nineteenth-century gastronomic writing, including Brillat-Savarin's book. The presence of this anecdote makes the desensualized dullness of the rest of the text even more apparent.

Toward the end of the book, Brillat-Savarin included "The Choice of Sciences," a poem of his own.[78] The voice of this poem renounces history, astronomy, chemistry and physics in favor of gastronomy, cookery and love. Here Brillat-Savarin presents himself like the gourmand in the frontispiece of the first volume of Grimod's *Almanac*. He seems disillusioned with the sciences even while trying to have gastronomy be recognized as one. His impulse to include the varieties in his book, even though they did not fit his chosen scientific format, might be an indication of his awareness of the serious limitations of gastronomy as a science. Brillat-Savarin's *Physiology*, and gastronomy in general, treat the sensual and affective aspects of taste as an excess that writers reluctantly had to tuck away.

The gastronomic writing of the nineteenth century is best understood as a site of negotiation in which those interested in food and taste as a field of practical and pleasurable knowledge tackled the philosophical and moral objections to their endeavor. The result was a compromise in which taste was forced into a straitjacket defined by rationalism and aesthetics. Gastronomy, by creating a different field for the study of taste, segregated writing about food. While gastronomy imitated the other disciplines, the other disciplines did not show any interest in learning from gastronomy. The fact is that, because gastronomy aped other disciplines rather than develop a new one based on the specificities of taste, it had very little to offer to established fields other than calling attention to how they had neglected gastronomy's subject matter. Whereas that in itself was a positive step toward the valorization of taste in Europe, it also stifled its sensuous and affective aspects. The gastronomic compromise was a pyrrhic victory, and gastronomers expressed their dissatisfaction with humor.

Notes

1 Carolyn Korsmeyer, "Tastes and Pleasures," *Romantic Gastronomies*, ed. Denise Gigante. January 2007. Par. 3. Romantic Circles. http://www.rc.umd.edu/praxis/gastronomy/korsmeyer/korsmeyer_essay.html. Accessed June, 2022.
2 Terry Eagleton, *The Ideology of the Aesthetic* (Oxford; Cambridge, MA: Blackwell, 1990), 15.
3 Eagleton, 16.
4 For an extensive discussion of the understanding of the sense of taste in Western philosophy see Carolyn Korsmeyer, *Making Sense of Taste: Food & Philosophy* (Ithaca, NY: Cornell University Press, 1999).
5 Denise Gigante, *Gusto: Essential Writings in Nineteenth-Century Gastronomy* (New York: Routledge, 2005), xxii.
6 For an analysis of food and taste in Romantic literature see Jocelyne Kolb, *The Ambiguity of Taste: Freedom and Food in European Romanticism* (Ann Arbor, MI: University of Michigan Press, 1995).
7 Sean R. Silver, "Locke's Pineapple and the History of Taste," *The Eighteenth Century* 49, no. 1 (2008): 50, https://doi.org/10.1353/ecy.0.0004.
8 Silver, "Locke's Pineapple."
9 François Marin, *Les Dons de Comus, ou l'Art de la cuisine, réduit en pratique*, 3rd ed., vol. 1 (Paris: Chez Pissot, 1758).
10 Marin, 1:xlvii. My translation.
11 Marin, 1:xlviii.
12 E.C. Spary, *Eating the Enlightenment: Food and the Sciences in Paris, 1670–1760* (Chicago: University of Chicago Press, 2014), loc. 3956–3967 of 9350, Kindle.
13 Charles Fourier, *The Utopian Vision of Charles Fourier: Selected Texts on Work, Love, and Passionate Attraction* (Boston: Beacon Press, 1971), 267–270.
14 Benjamin Bourcier, "Jeremy Bentham's Principle of Utility and Taste," in *Bentham and the Arts*, eds. Anthony Julius, Malcolm Quinn, and Philip Schofield (London: UCL Press, 2020), 230–231.
15 Philip Schofield, "The Epicurean Universe of Jeremy Bentham," in *Bentham and the Arts*, eds. Philip Schofield, Anthony Julius, and Malcolm Quinn (London: UCL Press, 2020), 40.
16 Charles Fourier, *Selections from the Works of Fourier* (New York: Gordon Press, 1972), 65.
17 Fourier, *Selections*, 64.

18 Quoted in Daniel Sipe, "Social Gastronomy: Fourier and Brillat-Savarin," *French Cultural Studies* 20, no. 3 (August 2009): 220, https://doi.org/10.1177/0957155809105744. My translation.
19 Fourier, *Selections*, 63.
20 Fourier, 62–63.
21 Fourier, 74–75.
22 Jean-Jacques Rousseau, *Emile: Or, On Education* (New York: Basic Books, 1979), 153.
23 Michel Foucault, *The History of Sexuality: An Introduction* (New York: Vintage Books, 1990), 58.
24 Spary, *Eating the Enlightenment*, 17–21.
25 Joseph Berchoux, *La Gastronomie*, 5th ed. (Paris: L.G. Michaud, 1819), 20. My translation.
26 Charles-Louis Cadet de Gassicourt, *Cours gastronomique ou les Diners de Manant-Ville/Ouvrage anecdotique, philosophique et littéraire*, 2nd ed. (Paris: Capelle et Renand, 1809), 294.
27 Jean-Baptiste Gouriet, *L'Antigastronomie, ou l'Homme de ville sortant de table, Poëme en IV chants....* (Paris: Chez Hubert, 1804), 23–24.
28 William Kitchiner, *The Cook's Oracle and Housekeeper's Manual* (New York: J. & J. Harper, 1830), xi.
29 Jean Anthelme Brillat-Savarin, *The Physiology of Taste, or, Meditations on Transcendental Gastronomy*, trans. M.F.K. Fisher (New York: Vintage Books, 2011), 30–32.
30 Thomas Walker, *Aristology, or the Art of Dining* (London: George Bell and Sons, 1881), 8–9.
31 Brillat-Savarin, 61.
32 Alexandre-Balthazar-Laurent Grimod de la Reynière, *Almanach des Gourmands: Servant de guide dans les moyens de faire excellente chère*, 2nd ed., vol. 3 (Paris: Chez Maradan, 1806), 1–7.
33 Brillat-Savarin, 155.
34 Brillat-Savarin, 241.
35 Brillat-Savarin, 246.
36 Brillat-Savarin, 254–260.
37 Brillat-Savarin, 54.
38 Spary, *Eating the Enlightenment*, 50.
39 Spary, 43.
40 Jessica Riskin, "The Defecating Duck, or, the Ambiguous Origins of Artificial Life," *Critical Inquiry* 29, no. 4 (June 2003): 599, https://doi.org/10.1086/377722.
41 Peter Dear, "A Mechanical Microcosm: Bodily Passions, Good Manners, and Cartesian Mechanism," in *Science Incarnate: Historical Embodiments of Natural Knowledge*, ed. Christopher Lawrence and Steven Shapin (Chicago, IL: The University of Chicago Press, 1998), 53.
42 Dear, 74–75.
43 Dear, 63.
44 Dear, 53.
45 Foucault, 1–35.
46 Berchoux, 35–36.
47 Grimod de la Reynière, *Almanach des Gourmands*, vol. 1 (Paris: Chez Maradan, 1803), 225–229. All translations of quotes from Grimod de la Reynière's work are my own.
48 Grimod de la Reynière, *Almanach des Gourmands*, vol. 6 (Paris: Chez Maradan, 1808), 158.
49 Brillat-Savarin, 160.
50 Emma C. Spary, *Feeding France: New Sciences of Food, 1760–1815* (Cambridge: Cambridge University Press, 2014), 250.
51 René de Chazet et al., *L'École des Gourmands* (Paris: Chez Mad. Cavanagh, 1804), 26.
52 Brillat-Savarin, 37–41.
53 Grimod de la Reynière, *Almanach des Gourmands*, 2nd ed., vol. 2 (Paris: Maradan, 1805), 93–104.

54 Brillat-Savarin, 45.
55 Julio Cortázar, *Cronopios and Famas*, trans. Paul Blackburn (New York: New Directions Pub. Corp, 1999), 21–22.
56 Roland Barthes, *The Rustle of Language* (Berkeley: University of California Press, 1989), 257.
57 Grimod de La Reynière, *Almanach des Gourmands*, 1803, 1:i–ii.
58 Grimod de La Reynière, *Almanach des Gourmands*, 1803, 1:ii–iii.
59 Grimod de La Reynière, *Almanach des Gourmands*, 1803, 1:iii.
60 Grimod de La Reynière, *Almanach des Gourmands*, 1803, 1:iii. All translations of quotes from Grimod's work are my own.
61 Grimod de La Reynière, *Almanach des Gourmands*, 1808, 6:66.
62 Grimod de La Reynière, *Almanach des Gourmands*, vol. 5 (Paris: Chez Maradan, 1807), 239–245.
63 Grimod de La Reynière, *Almanach des Gourmands*, 1808, 6:67.
64 Grimod de La Reynière, *Le Gastronome français, ou l'Art de bien vivre* (Paris: Charles Béchet, 1828), 45–48.
65 Grimod de La Reynière, *Le Gastronome français*, 477.
66 Anonymous, "Le Cerveau et l'Estomac," *Le Gastronome: Journal universel du goût* 1, no. 2 (March 18, 1830): 2–4.
67 Launcelot Sturgeon, *Essays, Moral, Philosophical, and Stomachical, on the Important Science of Good-Living*, 2nd ed. (London: Whittaker, 1823).
68 Grimod de La Reynière, *Le Gastronome français, Manuel des Amphitryons* (Paris: Capelle et Renard, 1808).
69 Sturgeon, 50.
70 Sturgeon, 51.
71 Brillat-Savarin, 347.
72 Brillat-Savarin, 348.
73 Brillat-Savarin, 348.
74 Brillat-Savarin, 353–354.
75 Brillat-Savarin, 350–353.
76 Brillat-Savarin, 351.
77 Brillat-Savarin, 352.
78 Brillat-Savarin, 413–414.

References

Anonymous. "Le Cerveau et l'Estomac." *Le Gastronome: Journal universel du goût* 1, no. 2 (March 18, 1830): 2–4.
Barthes, Roland. *The Rustle of Language*. Berkeley: University of California Press, 1989.
Berchoux, J. *La Gastronomie*. 5th ed. Paris: L.G. Michaud, 1819.
Brillat-Savarin, Jean Anthelme. *The Physiology of Taste, or, Meditations on Transcendental Gastronomy*. Translated by M.F.K. Fisher. New York: Vintage Books, 2011.
Cadet de Gassicourt, Charles-Louis. *Cours gastronomique, ou les Diners de Manant-Ville/ Ouvrage anecdotique, philosophique et littéraire*. 2nd ed. Paris: Capelle et Renand, 1809.
Chazet, René de et al. *L'École des Gourmands*. Paris: Chez Mad. Cavanagh, 1804.
Cortázar, Julio. *Cronopios and Famas*. Translated by Paul Blackburn. New York: New Directions, 1999.
Dear, Peter. "A Mechanical Microcosm: Bodily Passions, Good Manners, and Cartesian Mechanism." In *Science Incarnate: Historical Embodiments of Natural Knowledge*. Edited by Christopher Lawrence and Steven Shapin, 51–82. Chicago, IL: The University of Chicago Press, 1998.
Eagleton, Terry. *The Ideology of the Aesthetic*. Oxford; Cambridge, MA: Blackwell, 1991.

Foucault, Michel. *The History of Sexuality: An Introduction.* New York: Vintage Books, 1990.

Fourier, Charles. *Selections from the Works of Fourier.* New York: Gordon Press, 1972.

———. *The Utopian Vision of Charles Fourier; Selected Texts on Work, Love, and Passionate Attraction.* Boston: Beacon Press, 1971.

Gigante, Denise, ed. *Gusto: Essential Writings in Nineteenth-Century Gastronomy.* New York: Routledge, 2005.

Gouriet, Jean-Baptiste. *L'Antigastronomie, ou l'Homme de ville sortant de tabl, Poëme en IV chants.…* Paris: Chez Hubert, 1804.

Grimod de la Reynière, Alexandre-Balthazar-Laurent. *Manuel des Amphitryons.* Paris: Capelle et Renard, 1808a.

———. *Almanach des Gourmands: Servant de guide dans les moyens de faire excellente chère.* Vol. 1. Paris: Chez Maradan, 1803.

———. *Almanach des Gourmands: Servant de guide dans les moyens de faire excellente chère.* 2nd ed. Vol. 3. Paris: Chez Maradan, 1806.

———. *Almanach des Gourmands: Servant de guide dans les moyens de faire excellente chère.* Vol. 5. Paris: Chez Maradan, 1807.

———. *Almanach des Gourmands: Servant de guide dans les moyens de faire excellente chère.* Vol. 6. Paris: Chez Maradan, 1808b.

———. *Almanach des Gourmands: Servant de guide dans les moyens de faire excellente chère.* 2nd ed. Vol. 2. Paris: Maradan, 1805.

Grimod de La Reynière, Alexandre-Balthazar-Laurent, et al. *Le Gastronome français, ou l'Art de bien vivre.* Paris: Charles Béchet, 1828.

Kitchiner, William. *The Cook's Oracle and Housekeeper's Manual.* New York: J. & J. Harper, 1830.

Kolb, Jocelyne. *The Ambiguity of Taste: Freedom and Food in European Romanticism.* Ann Arbor: University of Michigan Press, 1995.

Korsmeyer, Carolyn. *Making Sense of Taste: Food & Philosophy.* Ithaca, NY: Cornell University Press, 1999.

———. "Tastes and Pleasures." *Romantic Gastronomies.* Edited by Denise Gigante. January 2007. Par. 3. Romantic Circles. http://www.rc.umd.edu/praxis/gastronomy/korsmeyer/korsmeyer_essay.html. Accessed June, 2022.

Marin, François. *Les Dons de Comus, ou l'Art de la cuisine, réduit en pratique.* 3rd ed. Vol. 1. Paris: Chez Pissot, 1758.

Riskin, Jessica. "The Defecating Duck, or, the Ambiguous Origins of Artificial Life." *Critical Inquiry* 29, no. 4 (June 2003): 599–633. https://doi.org/10.1086/377722.

Rousseau, Jean-Jacques. *Emile: Or, On Education.* New York: Basic Books, 1979.

Silver, Sean R. "Locke's Pineapple and the History of Taste." *The Eighteenth Century* 49, no. 1 (2008): 43–65. https://doi.org/10.1353/ecy.0.0004.

Sipe, Daniel. "Social Gastronomy: Fourier and Brillat-Savarin." *French Cultural Studies* 20, no. 3 (August 2009): 219–236. https://doi.org/10.1177/0957155809105744.

Spary, E.C. *Eating the Enlightenment: Food and the Sciences in Paris, 1670–1760.* Chicago: University of Chicago Press, 2014a.

Spary, Emma C. *Feeding France: New Sciences of Food, 1760–1815.* Cambridge: Cambridge University Press, 2014b.

Sturgeon, Launcelot. *Essays, Moral, Philosophical, and Stomachical, on the Important Science of Good-Living.* 2nd ed. London: Whittaker, 1823.

Walker, Thomas. *Aristology, Or the Art of Dining.* London: George Bell and Sons, 1881.

3

BUREAUCRATIZING TASTE

Gastronomy initiated a process in which the experience of taste became increasingly bureaucratized. Gastronomic writers fashioned themselves as rational experts on taste and proposed their texts as codes of law to govern cooking and eating. Cooks and industrial food producers were expected to labor according to such laws, and eaters – if they aspired to be recognized as modern gourmands, and not simple gluttons – were urged to discipline their taste accordingly. Two of the most influential gastronomic writers of the nineteenth century were Alexandre-Balthazar-Laurent Grimod de la Reynière (1758–1837) and Jean Anthelme Brillat-Savarin (1755–1826). Grimod was a lawyer and Brillat-Savarin was a judge. They both used the language of law extensively to establish their authority as objective and universal. Grimod de la Reynière, credited with the invention of the genre of food criticism, hosted courts of taste. He encouraged food producers and merchants to submit their products to his tribunal for evaluation or "legitimation." A positive evaluation would warrant a mention in his *Almanach des Gourmands* (1803–1812), which would ensure commercial success. The publication of food judgments in the guise of objective knowledge was an integral part of the formal rationalization of taste performed by gastronomy. Gastronomic publications served as codes that helped to standardize not only recipes and food products but also how gastronomic subjects experienced them.

The frontispiece to the second volume of Grimod de la Reynière's *Almanach* illustrates a queue of people waiting to submit their food products to the gourmand, hoping that he will "legitimate" them with an endorsement (see Figure 3.1).[1]

In this illustration, the gourmand is represented as a bureaucrat, not as a sensuous eater. Instead of being seated at table, he is seated at his desk with pen and paper at hand. The copious foods around him are in the process of being evaluated and cataloged, not served and eaten. The bureaucrat of taste is the beneficiary of free

DOI: 10.4324/9781003331834-4

Les audiences d'un Gourmand.

Dunant del.   Grimod de la Reyniere inv.   Mariage Sc.

FIGURE 3.1 Frontispiece from Grimod de la Reynière's *Almanach des Gourmands*, showing a gourmand receiving foods submitted for his evaluation. Photo courtesy
of the Bibliothèque nationale de France.

fine food, but he acquired it under the pretension of being a rational judge of taste.
He is laboring more than enjoying in a sensuous way.

The specific kind of rationality that shaped gastronomy and helped it advance its
goal of legitimizing interest in food and taste as a valuable aspect of modern society
was one Max Weber called formal rationality. This specific type of rationality has
its origins in the modern Western industrialized world, and it is institutionalized in
large-scale structures like the bureaucracy, modern law and the capitalist economy.[2]
The formal rationality of gastronomy can be seen in its codifying impulse and in

the construction of the gastronomic writer as a professional expert that mediates the relationship between bodies and food through the use of the printed word. This mediation facilitated the circulation of food commodities by spurring appetite while bypassing the process of personal judgment and selection. Seen in this way, gastronomy is the precursor, instead of the antithesis, of the fast-food culture in which efficiency and profitability take priority over taste considerations.

The process of bureaucratizing taste was slow and always contested. It is another aspect of the gastronomic compromise discussed in Chapter 2, in which gastronomic writers stripped their notion of taste of its sensuous subjectivity and affective aspects in their bid to appease the longstanding philosophical and moral devaluation of the sense of taste as lacking epistemic value and leading to excess. Gastronomic texts and cookbooks show the advancing process of the bureaucratization of taste. They also show some resistance to a rigid understanding of cooking that does not allow for invention and artistry and an understanding of taste that stigmatizes taste preferences that deviate from the ones being codified as rational and modern.

An important function of nineteenth-century gastronomic writers as bureaucrats of taste was to facilitate a transition from the refinement and excess of the aristocracy to the relative restraint of the bourgeois, and from court patronage to the capitalist market. Gastronomic texts established the rules of a discipline that created gourmands as modern subjects. European gourmands were the first ones to submit to the canons of the gastronomic discipline, but this discipline slowly became naturalized as the commonsense approach to the sense of taste in middle-class sectors of modern capitalist societies. Bourgeois notions of taste and cuisine were based on the simplification of the sophisticated ideals of the nobility, which required a long apprenticeship. The bourgeois, although affluent, did not dedicate the same levels of wealth and leisure to the cultivation of taste as the aristocracy could.

Manuals of good manners and taste proliferated in the nineteenth century as a result of shifting class relations. They provided the rising bourgeoisie with easy-to-follow rules and with other useful information for social promotion.[3] According to Olivier Assouly, manuals described and detailed the appropriate conduct for all kinds of social situations in such a way that readers only had to repeat and simulate the prescribed gestures and attitudes.[4] Gastronomic writing belongs to this genre of easy-to-follow rules for the bourgeoisie. Gastronomic texts provided simple guidelines and the basic knowledge necessary to be considered a rule bound and therefore rational gourmand, rather than an unreflective and unrestrained glutton. The simplification and standardization of gustatory taste in the shape of a code of law came to be not only because of the aesthetic requirement of abiding by objective principles but also because it was essential to facilitate the circulation of food and taste experiences as commodities in the growing capitalist market. It helped new consumers navigate the markets full of new food commodities, including many that came from the colonies. Gastronomic texts also taught readers how to run kitchens and entertain with limited skill or knowledge and on a middle-range budget.

An important change from aristocratic taste culture to bourgeois gastronomy was the increased legitimacy of consumption. Gastronomic writers took as their responsibility the introduction of food novelties, which they portrayed as signs of human progress. The consumption of new food products is cast not as decadent sensuality but as the gastronomic equivalent of being up to date with scientific discoveries. The framing of the consumption of food and other commodities as part of the march of progress gave the interest in food the legitimacy that philosophy had withheld for centuries. Brillat-Savarin, credited as one of the founding figures of gastronomic writing, stated that his writing was fueled by "a fear of lagging behind the times."[5] Being knowledgeable about food was no longer a pastime for the rich and idle but a privileged responsibility imposed by modern times. For his part, Grimod de la Reynière included "new discoveries" in his annual catalog of the fine foods available in Paris. For Grimod, progress meant the constant emergence of new products, like mustard made with Champagne wine and chestnut flour from Lyon.[6] Constant innovation in the way of new products that create new needs and wants is of course fundamental for the development of the capitalist market. Gastronomy, as the taste culture of the bourgeois, is geared toward capitalist consumption. Capitalism was more than a context for the emergence of gastronomy: it was its driving force.

While for a long time beholden to old aristocratic standards, gastronomic writers slowly articulated the discourse that defined a distinctively modern capitalist approach to taste. In the history of the relationship between capitalism and art, it can be observed that the market gave modern art a relative freedom from the exigencies imposed by the church, aristocratic standards and other external forces.[7] However, the market imposed its own standardization and massification logic and demands. Eventually the art vanguards rebelled against the market and its bourgeois mediocrity, creating movements like art for art's sake. But the history of the relationship between the market and gastronomy, which proposed itself as an art of taste, played out in a different way. Gastronomic writing was anchored in the logic of capitalism from the very beginning, but it did not develop a critical distance. Romantic literary writers and avant-garde artists have used food and gustatory taste to explore the limits of reason and to explode bourgeois conventions.[8] But the gastronomic genre as such did not antagonize the system to which it owed the conditions for its own existence.

The logic of capitalism allowed gastronomy to assert consumption as a virtue. Ingesting food and acquiring commodities became sanctioned as signs of modern progress. If the philosophical justification for gastronomy allowed for interest in food and even a big appetite to be considered rational as long as the menus were chosen following standards of taste and medical directives to avoid indigestion and ill health, the capitalist justification was based on the understanding of abundance and newness as progress, and on the reduction of reason to capitalist standardization. In 1846, Eugène Briffault bragged about the fact that the whole world was involved in the feeding of Paris and suggested that foods from all kingdoms and from all parts of the natural world rushed to please Parisian mouths.[9] According to Briffault, the whole world was supplying Paris out of its free will and desire to please its superior eaters. He was proud of the benefits of colonial and imperial production. For this author and many others in France and England, the global order in which the whole world supplied them was a sign of their superiority, which they took to be a natural given rather than the result of exploitative and brutal relations of power. Because now appetite and the ability to satisfy it was seen as a mark of a superior civilization, Briffault could celebrate the insatiable appetite of Paris without shame. He gives detailed statistics of how many animals were slaughtered in 1844 and continues to marvel about how much of each specific major food item Parisians devoured in that year.[10] In gastronomic writing, gluttony is not shunned on moral grounds. Instead, the consumption of food and other commodities was fashioned as a modern duty.

Gastronomic publications openly celebrated consumption as modern progress. In the *Cours gastronomique*, published by Charles-Louis Cadet de Gassicourt in 1809, the character known as "Professor" assures the bourgeois man that he is tutoring in gastronomic matters that there is no reason to fear that a sumptuary law restricting consumption would be enacted. He argues that even though luxury could be corrupting, it is also useful because it stimulates the movement of money, spurs industry and favors the progress of the arts.[11] Echoing this idea in 1894, Chatillon-Plessis argued that staying up to date with the development of inventions that enrich life at table was both a pleasure and a duty.[12] The capitalist transformations of new wants into needs allowed for the stigma on consumption and gluttony to be softened. Capitalism, more than the grand epistemological pretensions of gastronomy, is what finally validated the pleasures of taste in modern Europe.

Not everybody was convinced of the innocence of capitalist consumption. In The *Knife and Fork* (1849), the English author known as the Alderman makes fun of the gourmands' sense of entitlement:

> Your fame, as epicures, extends from pole to pole; and so universal and instinctive is the knowledge of your love of turtle, that, if report speak truly, the fishermen have only to shout your names on the sea beach to attract an army of turtles all ready to turn themselves on their backs, without assistance from any one, and to die, singing after their own fashion, "For they (yourselves) are jolly good fellows!"[13]

This humorous excerpt makes the point later theorized by Karl Marx as commodity fetishism. This concept refers to the magical character that objects acquire when they take the form of a commodity.[14] A commodity hides the social relations of labor that produced it and thus appears as if it were independent from human action. The turtles offer themselves to the epicure with little or no labor by fishermen or intervention of the market. The same principle was at work in Briffault's idea that the whole world just happens to want to please Parisian eaters. The violence of the global networks of production and commerce is hidden from view in the celebration of the cornucopia enjoyed at the centers of European empires. The celebration of the abundance that made gastronomy possible does not even acknowledge the suffering and hunger that capitalism and imperial colonialism created around the world in order to provide such unprecedented plenty for Europe.

Capitalism gave the sense of taste a relative freedom vis-a-vis moral and aesthetic injunctions against excess, but this should not be confused with an encouragement of the refinement of the sense of taste, or with a more complete enjoyment of its pleasures. Marx famously argued, "The cultivation of the senses is the work of all previous history."[15] He understood that the capabilities of the senses are not a natural given, but a historical and social construction. For Marx, the brutalization of the senses is one of the clearest examples of the dehumanizing character of capitalism. In industrial capitalism, the high temperatures, noise and dust of the factories injured the sense organs of the workers.[16] In the 1860s, for example, cotton manufacturers experimented with changes in the production process of cotton. These experiments, Marx tells us, "were not made just at the expense of the worker's means of subsistence. His five senses also had to pay the penalty."[17] The intolerable smell, dust and dirt of the bales of cotton irritated the skin, throat and lungs of the workers, leading to disease. The industrial capitalist worker's body and senses were under constant attack. Given their work and life conditions, the workers couldn't afford to develop and refine their senses. Doing so would only have brought them more suffering.

The bourgeois could escape the sensorial assault endured by the workers, but according to Marx they did not necessarily cultivate and gratify their senses either. The bourgeois developed an ethics of restraint that would allow them to accumulate wealth. As Marx sharply put it: "*all* the physical and intellectual senses have been replaced by the simple alienation of *all* the senses; the sense of *having*."[18] Because of private property, "an object is only ours when we have it, when it exists for us as capital, or when it is directly eaten, drunk, worn, inhabited, etc., in short, *utilized* in some way."[19] The relationship with the objects of sensation is limited to the affirmation of possession through the act of consumption. There is no engagement with the senses in a properly human, social or aesthetic way. The attacked and ungratified senses in industrial capitalism are a far cry from the fully human senses that Marx envisioned only the abolition of private property could bring about.

Gourmands, even while rejecting the ethics of restraint that Marx argued defines the bourgeois as a class, were certainly more concerned with having, in the sense of ingesting and digesting, than with exploring the experience of taste. Gourmands were excited about the abundance of old and new food products deemed excellent by the bureaucrats of taste. The gastronomic taste for abundance and novelty matches the credo of capitalist production. Both gastronomy and capitalism grew with their hunger for increased production and constant innovation. In this scenario, gastronomy initiated a process of standardization of cooking and eating. The foods that were declared most desirable by the taste bureaucracy were fossilized in cookbooks that were presented as codes of law, and eaters just had to read gastronomic texts to align their taste preferences with the gastronomic standard. As new food commodities were produced, they were inserted into the gastronomic matrix of codes, in a constant cycle of innovation and codification. Gourmands did not need to ask themselves what they liked. They asked instead: what am I expected to like next? The sense of taste was placed on the treadmill of progress understood as capitalist development. This is hardly a context conducive to pausing to explore the complexities of sensory experience. Consuming the latest commodities approved by the taste bureaucrats, more than gustatory refinement, is what defines the gastronomic notion of taste. The gastronomic standard of taste was not the result of the formulation of principles that could be considered objective by any definition. It was just the imposition and naturalization of the arbitrary and contingent preferences of a small elite, who gave no explanations or criteria for their judgments.

The symbiotic relationship between gastronomy and capitalism continued to develop. The weekly paper *The Table*, published by A.B. Marshall in England from 1886 to 1918, provides a window into how modern taste became increasingly ruled by the capitalist market. Ever since the beginnings of the gastronomic discipline with Grimod de la Reynière, it was clear that gastronomy was about guiding the bourgeois in their process of consumption. Grimod's "legitimations" were really an elaborate way of making personal endorsements that served as advertisements. In Marshall's *The Table*, we see plenty of straightforward advertisements of foods, culinary equipment and other products suitable for the bourgeois lifestyle. Marshall published her own recipes and advertised her culinary school, her food products and her inventions, including an ice-cream machine. Marshall was a cook, a gastronome and an entrepreneur all in one. As a businessperson she "legitimated" her own food products, bypassing the external review. The market increasingly co-opted and eroded the role of mediators of taste that the aristocrats had carved for themselves. Today the market is the leading tastemaker, as discussed in the conclusion of this book.

The development of modern taste through gastronomy was driven by the emerging global market. Gastronomers served as tastemakers that advertised the food commodities and made the gourmands receptive to them. What distinguished the gourmand as a modern subject was his participation in the capitalist market as a consumer. The sense of taste acquired the economic value of spurring the economy, thereby allowing for the status of taste to be raised. But it also standardized the sense of taste of the modern subjects of gastronomy, turning it into a passive interface between bodies and the market. Gastronomy constructed standardized bodies optimized for the consumption of standardized foods. Print was the main technology employed by gastronomy for the standardization of cooking and eating.

Print as a Technology of Taste

Gastronomic writing could be seen as an example of what Benedict Anderson described as print capitalism and its importance in the creation of the imagined community that is the nation.[20] Gastronomic scholarship has noted that gastronomic texts were an important element for the building of French national identity. But beyond helping to construct the imagined community of France, gastronomic texts founded the transnational imagined community of modern gastronomic subjects. Gastronomy used print for the dissemination of its ideas. Print was the technology used to teach modern middle-class subjects how to taste without physically tasting.

Given the authority bestowed on the printed word, print was an effective tool for the transformation of arbitrary notions of taste into universal rules. Print was also essential for the desensualization and standardization of taste, as gourmands learned to taste by reading instead of by tasting. In the modern period, new technologies of vision and hearing could be used to enhance sensual experience, but no significant technologies of taste and smell have been developed. In the absence of taste technologies, gastronomy used print as a technology of taste in a way that impoverished the taste experience by overruling and aiming to replace embodied taste.

Taste and smell, senses that were devalued in Western cultures because of their supposedly limited cognitive capabilities, have not been seriously challenged by technological devices capable of storing, reproducing and transmitting sense data. Sarah Danius argued that technologies like radiography, photography, phonography and telephony appropriated the epistemic privileges of the human senses, particularly vision and hearing.[21] She explains that technologies initiated new forms of knowledge and created a split between scientific and bodily vision and hearing. Human vision became discredited as a conveyor of truth, given that technological devices could see better. Technologies of sight and hearing redefined these senses and ultimately freed them from the epistemic burden that used to be the key to their value. In the mid-nineteenth century, the artist's eye claimed autonomy from the previously established rational and instrumental conventions of aesthetic perception.[22] Vision and hearing were favored over taste, smell and touch by aesthetic discourse because they were thought to produce objective knowledge. Modernist art is the result of sight and hearing being overruled, enhanced and transformed by technologies, thus freeing artists to explore their subjective perceptions. Modernist art emerged only after technologies relieved vision and hearing from their epistemic duties. There was no longer a need for realist painting, making surrealism and impressionism possible. In contrast, no taste and smell technologies comparable to telescopes, microscopes, X-rays and photography (not to mention cinematography) have ever emerged. There are no technologies capable of copying, storing, reproducing or enhancing a particular taste experience. Ironically, even though the specific cognitive abilities of taste and smell are not highly esteemed, no technology can do what they do any better than them.

The absence of major taste and smell technologies, and the low epistemic value that Western thought gave to these senses, made gastronomy tie its notion of taste to the search for truth and objective knowledge. Placing an epistemic burden on taste was a strategy used by gastronomic writers to legitimize gustatory pleasure. Lack of epistemic value was not liberating in the case of taste because this sense was also considered dangerous, as it can lead to excesses that corrupt the body and society. The sense of taste was cognitively irrelevant, but it still had to be subjected to strict control. Insisting on the cognitive qualities of taste was the way in which taste could overcome this moral stigma. A taste culture with a focus on pleasure and which stresses subjectivity as the key to artistic culinary practice could not arise in this context.

In the early modern period, there was a backlash against rich and highly seasoned dishes because they fooled the sense of taste into eating things that are not healthy.

Because tastiness does not equal healthfulness, critics of cuisine urged people to eat simple preparations that allowed the appreciation of the natural taste of the food.[23] The meaning of natural taste is not stable, as it has meant something different to different people in different times and places. In Europe, the ideal of the natural taste has referred both to plainly prepared foods and to foods that have been prepared in a way that brings out the natural taste of the food, understood as a hidden truth that is uncovered by proper cooking. In either case, the sense of taste was considered to be unable to taste properly on its own, without the bureaucratic mediation of doctors, cooks or gastronomers to steer eaters away from their gustatory preferences and toward food rationally chosen to help digestion and preserve health.

According to E.C. Spary, before the formalization of gastronomy, the figure of the gourmet had some epistemic legitimacy as a skilled wine connoisseur.[24] Many distillers were happy to depend on a connoisseur's palate to measure the qualities of wine and liqueurs, but the work of chemists like Antoine Baumé eroded the value of corporeal skill. In 1768, Baumé's invention of an aerometer, a device for measuring the spirit content of liqueurs by specific weight, was part of the effort of substituting calibrated laboratory instruments for the gourmet's body.[25] The aerometer was only partially successful, and no device has managed to totally overrule the role of human tasters in the assessment of goods like wine and tea. But the history of the aerometer demonstrates a will to bypass the body in favor of the development of technologies of taste and smell. The gastronomic notion of taste favored a disembodied experience that was believed to give access to objective knowledge over an embodied experience that is subjective and resists standardization. Print was the technology that helped gastronomy shape taste to fit this notion. Dependence on the printed word gave even more power to the desensualizing impulse of gastronomy. Gastronomic writing shifted the focus of eating from the mouth and the nose to the eyes as the medium of the disembodied mind. Writing and reading rules and guides are the defining activities of the gastronomic subject, who needed to distance himself from the fully embodied experience of taste to claim eating as a rational activity.

The development of taste technologies would not have freed taste from an epistemic function that it was supposed to not have to begin with. Instead, the lack of taste technologies became what enabled taste to claim a measure of epistemic value. This would not have been possible without the use of print to disseminate gastronomic texts that served as a supplement for taste. Brillat-Savarin was convinced that the art of printing was an insurance against the dangers of retrogression in all the sciences, including gastronomy.[26] Printing certainly helped gastronomy set its views and transmit them to a wide audience. Print, like the spoken word, can help people taste better by giving guidance on tasting criteria. It can also somehow recreate a taste by storing a description for a long time and disseminating it across space and time. However, this was not how gastronomic writing used the printed word. Gastronomy, defined by Brillat-Savarin as "the intelligent knowledge of whatever concerns man's nourishment,"[27] could not devote much time to sensuous descriptions of taste and pleasure that exceeded its mission of establishing objective food knowledge. Gastronomy used print to standardize the taste experience with rules, not to enable the exploration of the wide range of possibilities of the experience of taste. Even though there are many instances in which specific taste experiences and pleasures are conveyed in gastronomic writing, words whether printed or not, are a poor substitute for the fully embodied experience of taste. Matter of fact language, as Brian Massumi put it, dampens the intensity of affect.[28] The mediation of words and the scientific pretensions of gastronomy turned taste into a desensualized activity guided by predetermined rules.

The use of the printed word helped gastronomy establish its authority far and wide. Educated readers have been inclined to believe that the printed word gives privileged access to knowledge. The readers of gastronomic texts were expected to see them as bearers of the truth about taste. The fact that anything printed instantly had more authority facilitated the imposition of gastronomic rules. The power of print gave arbitrary ideas about taste the capability of being received as more authoritative than others, including the ones that could be derived from embodied tasting. The truth of taste was to be found in printed discourse, not in tasting. Arbitrary taste preferences were transformed into objective truths by virtue of the modern bias in favor of the printed word.

Brillat-Savarin was aware of the importance of sensory technologies for those interested in the pleasures provided by the senses. He praised lenses, telescopes and microscopes for extending the capabilities of the eyes, and mechanical technologies for fortifying the power of touch.[29] He lamented that civilization has done almost nothing to improve the other senses, but he finds a few important improvements: "The last few centuries have also brought important enlargements to the sphere of taste: the discoveries of sugar and its many uses, of alcohol, of ices, vanilla, tea, and coffee, have transmitted previously unknown pleasure to our palates."[30] For Brillat-Savarin, the impact of foreign and colonial products and their processing technologies are the most significant developments in the sphere of taste in centuries. This highlights how much the development of gastronomy owes to colonialism. Even though Brillat-Savarin did not think this colonial debt merited any further reflection, his words allow us to see the importance of colonial products for gastronomy. The colonial and capitalist supply of new foods was crucial for gastronomy to convey a sense of progress in the enhancement of taste.

Gastronomy also thrived with the introduction of new food preparation technologies, which gave cooks and gastronomers the feeling that cuisine and gastronomy were a part of the modern scientific march of progress. Ever since Grimod's first volume of the *Almanach des Gourmands*, gastronomic texts have included lists of "new discoveries" in which they celebrate technologies like canning and electricity together with new commercially available prepared mustards and vinegars. New technologies and new commodities were equally important for gastronomers who liked to see all their particularities as signs of universal progress. While gastronomers wrote about food and food technologies, I would argue that print was the most powerful technology in the shaping of the gastronomic approach to taste.

Gastronomic texts, as a mediation between foods and eating bodies, served to control pleasure by limiting the possibilities and the affective impact of the experience of taste. The pleasures afforded by gastronomy did not come from embodied sensory enjoyment but from compliance with gastronomic discipline, which was a way of asserting belonging in the class of modern cosmopolitan middle-class consumers. Readers of gastronomic texts did not need to use their sense of taste, they just had to spend their money buying the food commodities that confirmed their privileged position in the global capitalist order. The printed word was the technology that facilitated the creation of a taste for standardized foods and food novelties that can only be satisfied by the capitalist market and the modern/colonial global order.

Standardizing Cooking

The process of elevating the low epistemic status in which cooking and eating had been languishing in Europe for centuries was initiated by cooks as early as the seventeenth century. Cookbook authors demonstrated a great desire to raise their own social status by declaring themselves as the exclusive bearers of a useful knowledge. They aspired not only to spell out a system of cooking but to have it recognized as totally new and at the forefront of civilizational progress. Standardizing culinary practice was approached as the way to demonstrate that cooking was based on objective scientific knowledge. The preface of François Pierre La Varenne's *Le Cuisinier françois*, first published in 1651, explains that his intention with the publication of the book was to be helpful to those that might be in need, thus framing his knowledge as addressing a social need.[31] His cookbook has neatly classified sections, each one with its own table of contents. This book kept practical usability in mind, and many subsequent authors improved the strategies for presenting information in a user-friendly format. In 1674, L.S.R. published *L'Art de bien traiter*, a book that illustrates how cookbook authors made use of the modern belief in constant human progress as a way of establishing their own authority. L.S.R. confidently claimed that his book shows the "true science" of properly preparing and serving all kinds of food "with a method which has not yet been seen or taught, and which destroys all preceding ones as abusive, obscure and difficult to execute."[32] L.S.R. directly criticizes La Varenne as representing the kind of overwrought cooking that he considered a thing of the past.[33]

In their quest for collective and individual respect and glory, cooks inserted the idea of culinary progress into the larger colonial discourse of the civilizational superiority of Europe in general, and France in particular. François Massialot, in his 1691 *Le Cuisinier roïal et bourgeois*, argues that Europe is the only place where good taste reigns and where justice is done to the foods imported from other nations.[34] To his mind, in France they knew how to treat the alimentary wealth of the planet better than anywhere else. Massialot presented French cooks as custodians of a valuable knowledge that was evidence of the cultural superiority of Europe. Massialot's book provided an alphabetical index of recipes, which was another step toward the systematization of culinary information. Seventeenth-century cookbook authors in general strived to make culinary knowledge more accessible while at the same time assigning themselves the position of leaders in Europe's triumphant civilizational march forward.

In the changing philosophical and political climate of the eighteenth century, cooks saw an opportunity to shift the power balance between them and the people they served. Throughout the Enlightenment, cooks were invested in the idea of cooking as a science in a bid to transform themselves from servants into respected professionals. With the invention of *la cuisine moderne*, cooks claimed that their work required the use of scientific and medical knowledge.[35] Cooks wanted to be seen as more than manual laborers; they wanted to be recognized as rational workers whose knowledge was important for the health of their employers. The results of this strategy were mixed. As Sean Takats put it, cooks convinced others that their work involved medical and chemical principles, but they couldn't persuade the public that they were the ones who should handle this responsibility.[36]

Cookbook prefaces often rehearsed a history of cookery that placed the cookbook that they were introducing at the pinnacle of culinary progress. Cookbook authors routinely claimed that their work was entirely new, yet they often accused each other of plagiarism. Vincent La Chapelle described his book *Le Cuisinier moderne* (1735) as teaching to prepare all kinds of meals "in a more delicate way than what has been written so far," and he explains that the culinary changes of the past few decades are such that a table served in the style of 20 years ago will no longer satisfy guests.[37] He specifically decries the translation into English of Massialot's *Le Cuisinier roïal et bourgeois*, indicating that there has not been a new edition of that book in 30 years and that its contents are outdated.[38] However, Massialot accused La Chapelle of plagiarizing his work, and it has been shown that a third of the recipes in the first edition of *Le Cuisinier moderne* are copied.[39] The coexistence of strident claims of originality, on the one hand, and rampant plagiarism, on the other, indicates that there was a great desire for any kind of newness that could be presented as progress and that the claims of progress were more important than the actual pace and degree of culinary change. Writers and readers wanted to believe that they lived at the forefront of culinary and human progress.

For cooks to fashion themselves as rational thinkers, it was important to ascertain rules. The establishment of rules was crucial for the construction of cooking as an art. Cookbook authors negotiated a delicate balance between rules and innovation. They understood the importance of rules to establish their authority, but they still wanted the freedom to innovate. They did not want to be reduced to being mere bureaucrats. La Chapelle, who so proudly claimed to have presented totally new creations, stressed the importance of rules with the caveat that rules change and that "a Cook of Genius will invent new Delicacies."[40] In the fifth volume of the French edition of *Le Cuisinier moderne*, published in 1742, La Chapelle argued that rules are needed to slow down the irrational invention of cooks always in search of novelty.[41] In La Chapelle's mind, rules are fixed but they could be changed, although only by genius, rational cooks. Cookbook authors saw themselves as the vanguard of their profession and were as interested in lifting the social status of their class of laborers as in establishing their own individual genius and authority. Cookbook authors spurred the mythologization of cooks who are reputed to have single-handedly revolutionized cookery every few years.

Cookbook authors also developed arguments to overrule the taste preferences of their employers. Many cooks were increasingly employed in the homes of affluent bourgeois who lacked the fine dining know-how of the aristocracy. Cooks who had expertise in the taste standards of the aristocracy felt a sense of superiority over their new employers. Because they saw themselves as superior in knowledge and sensibility, they expected their employers to submit to their expertise. In the preface to the 1753 edition of Menon's *La Cuisinière bourgeoise*, the author disdainfully explains that he is not writing for nobles, but for the bourgeois, and that the book therefore presents preparations that do not require skill from the cook or opulence from their masters.[42] The female cook of the book's title is seen as lacking skill, and the bourgeois is dismissed as a parvenu. The idea of the objectivity of taste was useful in the cooks' quest to impose their authority over their employers. If an employer did not approve of what the cook provided, it was the fault of his inferior bourgeois taste.

In *La Science du maître d'hôtel, cuisinier* (1768), Menon included a "preliminary dissertation on modern cuisine." In this little essay, the author talks about cuisine as an art and about the cook as an artist whose work can only be appreciated by those with a "sagacity of taste."[43] He goes on to explain that the work of highly skilled artists often does not satisfy the common taste.[44] The clear message is that if you do not like his work it is because your taste is not sophisticated, which puts you in the company of the common people or of the "savage peoples of America or Africa," who would be unmoved by the music of the best of musicians.[45] The social surveillance function of modern taste was clear: conform to the standards that the bureaucrats of taste have established or be cast as inferior. Gastronomy discouraged the exploration of the many different ways in which taste can be experienced.

The culinary changes that characterize the modern cuisine of the eighteenth century were driven by new medical theories associated with chemical physicians that conceptualized digestion as a process of fermentation. The author of the preface to *Les Dons de Comus* (1742) described modern cuisine as "a kind of chemistry."[46] As Rachel Laudan has explained, the dishes and techniques of this new style of cooking were based on the theme of refinement "in the chemical sense as the preparation of the most enhanced form, the essence."[47] The centrality of stock in French cuisine can be traced back to the cooks' application of the scientific principles of their time. Modern cooking did not liberate taste from medical strictures but rather submitted it to the changing paradigms of scientific and medical theories. Refined cooking was supposed to aid digestion, which backed up the idea of the importance of cooks and cookery beyond sensual satisfaction. The chemical "fineness" of modern cuisine has subsequently been understood as referring to a refined aesthetics that corresponds to a superior class and civilization. But chemical fineness made cooks obsess over osmazome, stocks and gelatin, trying to find the "truth" of food as nourishment. Gastronomy's understanding of fineness was about digestibility and nourishment, not about the refinement of pleasure. This begins the path that led to nutritionism and to the industrialization of food production. Gastronomy is in a continuum with these developments; it is not its antithesis as the mythologizers of gastronomy propose.

Nineteenth-century cookbook authors continued to look back at the court cuisine of the eighteenth century with utmost respect, but they defended their right and ability to make changes to keep up with the times. There was a continuous ambivalence between the need to establish their authority by following rules and the desire for freedom to innovate. Whereas some cooks were excited about changes that they saw as signs of progress, many others feared the changes were a sign of decadence. Antoine B. Beauvilliers, in *L'Art du cuisinier* (1814), considers that French cuisine reached such high levels of perfection in the previous century that it would be hard to surpass it.[48] He considers that he is authorized to set standards because of his 40 years of experience working in high society settings.[49] Beauvilliers was certain that the readers of his book would be pleased by his recipes because the favorable judgment of the marquis de la Voppalière on matters of taste assured their value.[50] The standards of taste that French cuisine and gastronomic writing imposed as universal can be traced back to the standards established by the French aristocracy. The proliferation of cookbooks, nevertheless, allowed for these standards to be mimicked by a larger audience of apprentices who saw themselves as a growing aristocracy of taste.

Marie-Antoine Carême (1784–1833), one of the most influential cooks of the nineteenth century, had first-hand knowledge of the highest culinary practices of France, England and Russia. He was immensely proud and self-assured about the value of his contributions to the continued progress of French cuisine. His classification of sauces was a decisive step toward the systematization and standardization of cooking, and his sugar sculptures are a clear example of a cuisine that paid increased attention to the presentation of food in its effort to be regarded as a visual art. Sugar sculptures were also a monument to European colonial imperialism, which made sugar abundant at the expense of the labor, knowledge and resources of enslaved and colonized peoples.

Carême was bitter about what the changing sociopolitical order of his times meant for cuisine and for the importance of cooks like himself. He considered revolutions to be disastrous for cuisine.[51] He resented that French cooks were well received in the capitals of Europe while those who stayed in Paris were not appreciated. He argued that the great culinary talents languished "as a consequence of the miserable economy of the rich men that do not deserve the title of noble amphitryon."[52] Another major resentment of Carême's was against the rising numbers of gastronomic writers who were not cooks. He dismissed these writers as compilers and takes the time to refute in detail a few of the claims that such writers had popularized. Most importantly, he wanted to challenge the idea that before Grimod de la Reynière worldly men would have been embarrassed to talk about cuisine.[53] Carême considered that the continued progress of cuisine and gourmandise was not the result of a revolution spurred by Grimod de la Reynière, but it was the product of the work of renowned contemporary cooks like himself. In spite of Carême's efforts, the expansion of markets and of the reading public meant that star chefs like him were not the only ones participating in the establishment of cooking and taste standards.

Other cooks were not as displeased with the changing social order as Carême was. His disciple Jules Gouffé (1807–1877) did not think that French cuisine was decaying and considered that the conditions of his time were conducive to culinary greatness.[54] Gouffé divided his book into two parts: one for high-class cooking and one for domestic cookery. In the first part, he promised to present the latest developments in the progress of the art, while in the second part he used the watch and the scales as guarantors of a precision that would allow anybody to become a competent domestic cook. At the same time, ordinary cooks organized unions and published magazines in which they tempered the euphoria of celebrating the progress and superiority of French cuisine with their demands to be educated and to have their fair share of food.[55] Indeed, in 1891 cooks lamented that France was behind because it lacked culinary schools that were already flourishing in the United States, England, Germany and Switzerland.[56] Simplification and standardization were needed for the consolidation and spread of modern cooking. Whereas elite French cooks and gastronomers fought to be regarded as the source of gastronomic knowledge, it was clear that gastronomy was developing with considerable and even leading input from international markets beyond their control.

The divulgation of cooking knowledge took on a wider scope than chefs like Carême anticipated or would have approved. As more people became knowledgeable of the rudiments of elite cooking, the value of cooks did not increase as much as they had hoped, except in the case of a few star chefs. Instead, the emergence of gastronomic writing solidified the hegemony of gastronomers as critics who became the most authoritative bureaucrats of the modern discipline of taste. Gastronomers mediated the relationship between food producers and consumers, and as such were extremely influential in defining and spreading the discipline of taste. Cooks, on their part, continued to labor and to exert their influence in restaurants. Restaurants, more than home entertaining, were the main disciplinary institution in which middle-class consumers were tested. They had to publicly perform and display their gastronomic expertise by adequately choosing from the menu following the standards set by gastronomers. Beyond France, the ability to read the menu in French added another layer to the social surveillance function of gastronomy and the restaurant, since failure to understand French exposed customers as not belonging in modern bourgeois society.

Restaurants accelerated the process of simplification and standardization of cooking. Before restaurants, the standardization of dishes and their names was not necessary.[57] The standardization of cooking took lasting shape in 1903 with the publication of Auguste Escoffier's *Le Guide culinaire*, which is dubbed as the bible of French high cuisine. Escoffier codified recipes and also created a system of working in the kitchen akin to Fordist assembly lines. He explained his system responded to the need to simplify preparations to adapt to the increased speed of life.[58] Escoffier was sure that his system did not compromise the excellence of French cuisine, but it could be argued that his system laid down the path that led to fast-food restaurants. The defining concerns of restaurant cooking are speed, replicability and cost control, not just gustatory excellence. Gastronomy, and the restaurant as its main disciplinary institution, formalized the taste logic of industrial capitalism. The printed word established the rules of cooking and eating, and the restaurant was the most prominent space where taste was deployed for the surveillance of membership in polite modern capitalist societies.

Standardizing Eating

If cooks took to the printed word to standardize recipes and cooking procedures, gastronomers used it to standardize eating and the taste experience. Because of their preoccupation with the establishment of the legitimacy of gastronomy as a field of knowledge, gastronomers dedicated themselves to the compilation of information that could be related to food in one way or another. Together with the focus on digestion, amassing food-related information was the way in which gastronomy was fashioned as a rational endeavor. Gastronomic texts represent a printed word turn in the approach to taste. To downplay the physical and subjective aspects of taste, which were incompatible with the prevailing notion of reason, gastronomy turned tasting into an activity thoroughly mediated by printed discourse. All taste cultures enrich the experience of taste with oral and/or written discourses about food. Gastronomy did not start anything new when it comes to enriching taste by relating it to time and place-specific notions of philosophy, arts and science. What gastronomy did start was the use of the printed word to standardize taste. Gastronomic taste was not achieved by exploring the complexities of the taste experience but by reading to learn what a gourmand is expected to like and dislike.

Texts written and non-written can both enhance or stifle sensory pleasure, depending on how they are employed. The use of the written word in the case of gastronomy alienated sensory pleasure. Gastronomy drowned tasting in words. Brillat-Savarin said that he thought that something more than a cookbook should be written about food.[59] Certainly, Grimod and the other gastronomers who wrote before Brillat-Savarin shared the same feeling. Once the idea that interest in food could be the subject of rational discourse, there was a new mine of information to be gathered. Since food touches all aspects of human existence, gastronomers set out to document all the possible ways in which food is important. Gastronomic texts are quite lengthy, but the bulk of the pages do not discuss eating and tasting. They are mostly collections of information related to food. Brillat-Savarin feared being considered a compiler, but indeed a compiler he was.

Many gastronomic writers, particularly British ones, were aware that their genre was nothing more than rearranging information. In *Apician Morsels* (1829), Dick Humelbergius Secundus bluntly declares: "Modern writers on diet have added very little to the store of general information. The best of them are mere theorists and inexperienced speculators, and for the most part servile copyists, detailing from month to month what has been vulgarly known for centuries."[60] William

Makepeace Thackeray is another author who ridiculed the proliferation of books on obvious matters. He pointed out that snobs existed for years and years but when people became aware of their existence his book on snobs became unavoidable.[61] It was not the case that people had never given food and taste thoughtful consideration. It was just that the food and taste knowledge that is taken for granted in any given culture was not necessarily collected in print volumes or even written. One of the main reasons why gastronomy has been taken to be the first time that serious attention was given to the art of cooking and eating is that it was thoroughly documented in print. Of course, just because the taste and culinary knowledge of other times and places are not recorded in print in a separate gastronomic genre, it should not be assumed that it did not exist or was less sophisticated.

The enthusiasm for writing and publishing what normally are the unwritten standards of taste has a lot to do with the changes brought about by the French revolution and the rise of capitalism. In social contexts where change is more gradual, there is no need to write to explain what people already know and practice. Gastronomic writing served to transfer the food and taste know-how of the French court to the new affluent class in France and beyond. In the *Manuel des Amphitryons* (1808), Grimod de la Reynière passed on practical information regarding hosting etiquette,[62] while in the *Almanach*, he greatly expanded his scope and inaugurated gastronomic writing as a genre that juxtaposes the contents, style and conventions of many different genres. Grimod above all established the point of view of gastronomic writing as a gourmand's eye view of the world. Aside from information directly related to food consumption, like describing food items, indicating their seasonality and telling readers where to buy the best products, Grimod's and subsequent gastronomic publications included bits of literature, history, politics and art. Gastronomers wanted to document the importance of eating as it relates to all aspects of human existence. After Grimod, gastronomic writing took off. There were books, journals, almanacs, dictionaries, and encyclopedias. Many of them had multiple editions and translations. There was a frenzy to harvest all the food-related bits of knowledge available and to document the many changes that they invariably portrayed as progress. Food had become something to write and read about, creating a new food-related pleasure that did not necessarily involve actual eating. However, gastronomic texts are a potpourri of randomly arranged bits of information, and the gastronomers were only dilettantes in the fields of knowledge that they were mining.

Since taste was not considered an acceptable interest in itself, gastronomers not only read and wrote about eating, they instructed their readers to talk while eating. According to Grimod, the gourmand was not only somebody with a great appetite but also someone whose memory was adorned by a multitude of anecdotes and amusing stories that he could tell between dinner courses so that the moderate diners would forgive his appetite.[63] Grimod suggests that good conversation is required to make the gourmand socially acceptable. By his account, talking while dining was not geared toward the enhancement of the taste experience but about atoning and focusing attention away from the physical act of eating.

Gastronomic writing used the power of print not only to document and transmit food-related knowledge but to give it the weight of a code of law. The power of written codes was no secret, as it had been key to the establishment of imperial and colonial rule. If the modern era could be said to start with the Spanish colonization of the Americas in 1492, it also started with the publication of the first grammar of a modern romance language in the same year. Nebrija, author of the grammar, stressed on its prologue that language has always accompanied empires.[64] He saw a standardized language as a guarantor of imperial power by controlling speech variations and fixing memory. Similarly, gastronomic writing served to control not only eating behavior but also the very possibilities of how taste is experienced. The printed word, particularly in a time when few people could become published authors, gave the very local taste standards of the French court and their subsequent transformation in the hands of a transnational capitalist class the weighty appearance of a universally valid and superior culture.

In the effort to shape gastronomy as an art or science, gastronomers dispensed their perspectives on food as if they were objective and universal truths. This gave gastronomy a rather authoritarian character. Gastronomic texts were supposed to be seen as the final word on their subject, or at least until new scientific or technological advances made updates necessary. As it was the case with cookbook authors, gastronomic writers often claimed to have updated their predecessors. Differences from one text to the next were not treated as differences in preference or opinion but as one more step in a linear march of progress. Ever since Joseph Berchoux's pioneering poem *La Gastronomie* (1805) addressed its readers as pupils ignorant of the laws of gastronomy,[65] gastronomic texts have addressed their readers as subjects in need of being taught. Gastronomic texts often included aphorisms, as easily quotable gastronomic truths. Grimod argued for the need of a gourmand journal to spread the best food doctrine.[66] The word "doctrine" suggests that there are foundational unquestionable tenets that must be written down for them to be accepted by anybody aspiring to a gourmand identity. Brillat-Savarin said that the two main purposes of his book were to set forth the basic theories of gastronomy and to define gourmandism as a social grace.[67] He saw his book as the foundational text of a social practice and himself as its founding lawgiver. He warned that when he took on the voice of "The Professor" the reader "must bow down."[68] However, he admitted that when not using the voice of the professor his ideas could be debated. While aspiring to be recognized as an authority, he was well aware that his subject could not be fully accounted for in an objective manner.

Surprisingly little in gastronomic writing is dedicated to teaching readers how to appreciate and explore taste. Gastronomic texts often list memorable menus without an explanation of what made them memorable. The menus are not accompanied by any kind of description. Given that readers are supposed to be students who know nothing, they were expected to take those menus as models of what they should like. They were told what is good, and they were expected to accept the gastronomers' judgment as their own. This way of learning about taste by reading and accepting the standards established by others was only possible because of their belief in the objectivity of taste, or rather, because of their willingness to treat taste as if it were objective, disregarding its obvious subjective aspects for the sake of appearing respectable.

The establishment of law is also the policing of deviance. While complying with gastronomic rules conferred bourgeois distinction, failure to comply was supposed to expose a physical, moral or cultural deficiency. Gastronomy is based on essentialist ideas about eaters and about food. Whereas the experience of taste is notoriously variable according to biological, cultural and psychological factors, gastronomy was built on the notion that taste is an objective quality of foods that does not depend on the eater in any way. Gastronomy also saw eaters as subjects who have a clear, bounded and unchanging character. Gastronomy conceived of taste as something static rather than as an experience or a relationship between a food and a consuming body. Taste conceived as an experience, as many taste cultures have conceived it before, during and after gastronomy, can have different results in different contexts because taste does not become fully defined until it is experienced. The experience of taste could also transform the eating subjects or reveal them in a different guise. In contrast with this affective notion of taste, gastronomic taste is a sterile bureaucratic transaction in which an eating subject just grasps an objective taste. Because gastronomy had such an essentialist and static understanding of taste, it was easily deployed as a tool to discover the "truth" about eaters.

The food "legitimations" performed by Grimod's tasting jury are a clear example of how gastronomic texts provided the codes for the management of taste. Gastronomers became critics whose infallible taste judgment mediated between producers and consumers. Grimod and his friends put on a real performance of their self-appointed roles as judges of taste. In the *Almanach*, Grimod describes how the tasting jury conducted the tasting sessions with all the formality and presumed impartiality of a court of law. He explains that the written record of the judgments acquires the full force of law after a period during which they could be revised.[69] Under official circumstances, similar to the Pope speaking ex cathedra, their judgment was deemed infallible. Everybody else had to take their word for it, for their privileged bodies were the warrantors of excellence. The gastronomic standards of taste were ultimately based on the experience and opinions of aristocratic bodies, immortalized as laws in print.

While Grimod's tribunal of taste was supposed to reveal the truth about the foods that were judged, Brillat-Savarin designed gastronomical tests to reveal the truth about eaters. For Brillat-Savarin, gastronomical tests were "dishes of recognized savor and of such acknowledged excellence that nothing more than the sight of them will awaken, in a well-balanced man, all his gustatory powers."[70] He writes of gastronomical tests as a discovery, not a creation, in keeping with the idea that gastronomy is a science that follows natural laws. In this case, what was meant to be discovered is whether an eater is a real gourmand or not. Dinner hosts should not have to waste their efforts on guests who are only pretending to be gourmands. The clues to sorting out the real gourmands were both physical and verbal. The body of the person being tested had to show visible signs of delight and desire at the sight of the dishes being served, and he should also praise the food out loud. Gourmand is a title that only a small tasting elite deserves, and the way to identify them was not only by talking and pretending to know a lot about food but by physical "gut" reactions. Even though gastronomy was meant to teach all how to be gourmands as rational epicures, the naturally gourmand elite was revealed by their bodies and not by their knowledge alone. The gastronomical tests recognize that taste is a bodily affective experience, but this was not compatible with what gastronomy could teach while staying inside its moderate philosophical parameters. The embodied and affective experience of taste thus became a privilege reserved for the taste bureaucrats. The rest had to be content with tasting through the mediation of printed texts and the judgment of the body of experts.

Brillat-Savarin adjusted the gastronomical tests by social class, designing different menus for those whose incomes defined their status as either mediocrity, ease or wealth. He acknowledged that taste varies by class, but he understood this in an essentialist way. A man of wealth was not expected to be enthusiastic about the food of an economically inferior class, and a man of mediocre income was expected to be unmoved by the more expensive foods that he could not afford. Class and other differences were taken as natural and unchanging qualities of bodies, not the result of embodied history susceptible to change. What the invention of gastronomical tests reveals is the power of gastronomy to serve as the creator and guardian of social hierarchies. Brillat-Savarin's most famous aphorism, "Tell me what you eat, and I shall tell you what you are,"[71] captures the essentialist thinking of gastronomy. Eating habits are supposed to reveal the true identity of an eater. Grimod's tasting jury and Brillat-Savarin's taste council seem to take Immanuel Kant's judgment of taste quite literally. A tasting elite in secret meetings constituted tribunals of taste that determined the judgment criteria and published their rulings as edicts. The readers of gastronomic texts submitted to their authority to learn how to behave and talk as the members of the taste, class and civilizational elite. Disciplining their taste in this way was one of the requirements to avoid being dismissed as belonging to a lesser class or as a lesser human that is behind in the scale of gastronomic progress.

Gastronomy functions as a policing discourse that could unmask and expel those unworthy of the title of gourmand, bourgeois and civilized. Gastronomic discourse wanted to reveal the truth about eating subjects and food objects, not unlike how the discourse of sexuality was keen on revealing the truth of sex, according to Michel Foucault.[72] The truth of taste was not a trivial matter, as it was used to classify and hierarchize peoples and cultures. The role of gastronomic knowledge in policing social hierarchies in the Parisian context was clearly illustrated by Eugène Briffault in *Paris à table* in 1846. Briffault argues that Boileau's 1165 satire *Le Repas ridicule* about a disastrous dinner seems kind when compared with the cruel sarcasms faced by those who make mistakes when entertaining. He narrates how guests at the house of a distinguished woman make nasty remarks about the lack of ice in the water, the lack of freshness of the fish and the bad quality of the champagne. The remarks are accompanied by uproarious laughter, and the guests finally leave declaring they are hungry and one of them invites the rest for supper at the Café

Anglais. Money and position did not guarantee respectability if the standards set by the bureaucracy of taste were not followed. The gourmands in this anecdote were certainly relishing their newfound power to humiliate others. Gastronomic subjects have enjoyed the social distinction afforded by gastronomy. However, this distinction came at the steep price of accepting a bureaucratized and non-affective relationship with their sense of taste.

Notes

1 Alexandre-Balthazar-Laurent Grimod de La Reynière, *Almanach des Gourmands: Servant de guide dans les moyens de faire excellente chère*, 2nd ed., Vol. 2 (Paris: Chez Maradan, 1805), frontispiece.
2 For a summary and elaboration of Weber's theory of rationalization see George Ritzer, "The Weberian Theory of Rationalization and the McDonaldization of Contemporary Society," in *Illuminating Social Life: Classical and Contemporary Theory Revisited*, ed. Peter Kivisto, 5th ed. (Thousand Oaks, CA: Pine Forge Press, 2011), 46.
3 Olivier Assouly, *Le Capitalisme esthétique: Essai sur l'industrialisation du goût* (Paris: Editions du Cerf, 2008).
4 Assouly, 47.
5 Jean Anthelme Brillat-Savarin. *The Physiology of Taste, or, Meditations on Transcendental Gastronomy*, trans. M.F.K. Fisher (New York: Vintage Books, 2011), 30.
6 Grimod de La Reynière. *Almanach des Gourmands*, 2nd ed. Revue, corrigée et augmentée, vol. 2 (Paris: Chez Maradan, 1805), 249–260.
7 For a discussion see for example Gilles Lipovetsky and Jean Serroy, *L'Esthétisation du monde: Vivre à l'âge du capitalisme artiste* (Paris: Gallimard, 2013).
8 For a discussion of how the avant-garde developed anti-gastronomies see Cecilia Novero, *Antidiets of the Avant-Garde: From Futurist Cooking to Eat Art* (Minneapolis: University of Minnesota Press, 2010).
9 Eugène Briffault, *Paris à table* (Paris: J. Hetzel, 1846), 3, http://catalog.hathitrust.org/Record/006546806.
10 Briffault, 5–6.
11 Charles-Louis Cadet de Gassicourt, *Cours gastronomique, ou, les Diners de Manant-Ville / Ouvrage anecdotique, philosophique et littéraire*, 2nd ed. (Paris: Capelle et Renand, 1809), 96.
12 Chatillon-Plessis, *La vie à table à la fin du XIXe siècle: Théorie, pratique et historique de gastronomie moderne* (Paris: Firmin-Didot, 1894), 19–20.
13 Alderman, *The Knife and Fork* (London: H. Hurst, 1849), 2–3.
14 Karl Marx, *Capital: A Critique of Political Economy*, trans. Ben Fowkes, vol. 1 (London: Penguin, 1976), 164–165.
15 Karl Marx, *Karl Marx Early Writings*, trans. T.B. Bottomore (New York: Toronto: London: McGraw-Hill, 1964), 161.
16 Marx, *Capital*, 552.
17 Marx, *Capital*, 1:586.
18 Marx, *Early Writings*, 159–160.
19 Marx, *Early Writings*, 159.
20 Benedict Anderson, *Imagined Communities: Reflections on the Origin and Spread of Nationalism*, Rev. and extended ed. (London; New York: Verso, 1991), 37–46.
21 Sara Danius, *The Senses of Modernism: Technology, Perception, and Aesthetics* (Ithaca, NY: Cornell University Press, 2002), 1–24.
22 Danius, 55.
23 Viktoria von Hoffmann, *From Gluttony to Enlightenment: The World of Taste in Early Modern Europe*, Studies in Sensory History (Urbana, IL: University of Illinois Press, 2016), loc. 1733–1775 of 7807, Kindle.

24 E.C. Spary, *Eating the Enlightenment: Food and the Sciences in Paris, 1670–1760*. (Chicago: University of Chicago Press, 2014), loc. 3007 of 9350, Kindle.

25 Spary, loc. 3024–3035 of 9350, Kindle.

26 Brillat-Savarin, 60.

27 Brillat-Savarin, 61.

28 Brian Massumi, "The Autonomy of Affect," *Cultural Critique* Autumn, no. 31 (1995): 86, https://doi.org/10.2307/1354446.

29 Brillat-Savarin, 39–40.

30 Brillat-Savarin, 41.

31 François Pierre La Varenne, *Le Cuisinier françois, Enseignant la manière de bien apprester et assaisonner toutes sortes de viandes...* (Paris: Chez Pierre David, 1651), iiij.

32 L.S.R. *L'Art de bien traiter* (Lyon: Marchand, 1693), ii. My translation.

33 L.S.R., 4–5.

34 François Massialot, *Le Cuisinier roïal et bourgeois*, new and expanded (Paris: Claude Prudhomme, 1705), iii.

35 Sean Takats, *The Expert Cook in Enlightenment France* (Baltimore: Johns Hopkins University Press, 2011).

36 Takats, 11.

37 Vincent La Chapelle, *Le Cuisinier moderne, qui aprend à donner toutes sortes de repas*, vol. 1 (La Haye: De Groot, 1735), 1–2. My translation.

38 La Chapelle, 1:2.

39 Edmond Neirinck and Jean-Pierre Poulain, *Historia de la cocina y de los cocineros: Técnicas culinarias y prácticas de mesa en Francia, de la Edad Media a nuestros días* (Barcelona: Editorial Zendrera Zariquiey, 2001), 47.

40 La Chapelle, *The Modern Cook* (London: Nicolas Prevost, 1733), i.

41 La Chapelle, *Le Cuisinier moderne*, ed. D Morcrette, Luzarches, 2nd ed., vol. 5 (La Haye, 1984), iii.

42 Menon. *La Cuisinière bourgeoise, suivie de l'Office a l'usage de tous ceux qui se mêlent de dépenses de maisons* (Bruxelles: Foppens, 1753), ii–iii.

43 Menon, *La Science du maitre d'hôtel cuisinier, avec des observations sur la connaissance et propri- etés des alimens* (Paris: Compagnie des Libraires Associés, 1768), v–vi.

44 Menon, *La Science*, vi.

45 Menon, *La Science*, vi.

46 François Marin, *Les Dons de Comus, ou, l'Art de la cuisine, réduit en pratique*, 3rd ed., vol. 1 (Paris: Chez Pissot, 1758), xxii.

47 Rachel Laudan, "A Kind of Chemistry," *Petits Propos Culinaires* 62 (1999): 19.

48 Antoine B. Beauvilliers, *L'Art du cuisinier* (Paris: Pilet, 1814), viii.

49 Beauvilliers, xi–xii.

50 Beauvilliers, v.

51 Marie-Antoine Carême, *L'Art de la cuisine française au XIXe siècle. Traité elementaire et pratique*, vol. II (Paris: Comptoir des Imprimeurs-Unis, 1847), ii.

52 Carême, *L'Art de la cuisine française*, II:viii. My translation.

53 Marie-Antoine Carême, *Le Cuisinier parisien, ou, l'Art de la cuisine française au dix-neuvième siècle.*, 3rd ed., vol. 1 (Paris: Renouard, 1842), 14–15.

54 Jules Gouffé, *Le Livre de cuisine* (Paris: Hachette, 1867), v.

55 An example is *La Science culinaire*, which was the official publication of the Union uni- verselle pour le progrès de l'art culinaire.

56 Union philanthropique culinaire et d'alimentation. *La Cuisine française et etrangère* 1, no. 1 (August 15, 1891): 1.

57 Rebecca L. Spang, *The Invention of the Restaurant: Paris and Modern Gastronomic Culture* (Cambridge, Mass. London: Harvard University Press, 2001), 192.

58 Auguste Escoffier, *Le Guide culinaire: aide-mémoire de cuisine pratique* (Paris: Flammarion, 2004), vi.

59 Brillat-Savarin, 30.

60 Dick Humelbergius Secundus, *Apician Morsels: Or, Tales of the Table, Kitchen, and Larder* (London: Whittaker, 1829), 6.

61 William Makepeace Thackeray, *The Book of Snobs* (London: Punch Office, 1848), 3.

62 Alexandre-Balthazar-Laurent Grimod de la Reynière, *Manuel des Amphitryons* (Paris: Capelle et Renard, 1808).

63 Grimod de la Reynière, *Almanach des Gourmands*, 2nd ed., vol. 2 (Paris: Chez Maradan, 1805), vii.

64 Antonio de Nebrija, "Prólogo," in *Gramática de la lengua castellana* (Salamanca, 1492).

65 Joseph Berchoux, *La Gastronomie* (Paris: Giguet et Michaud, 1805).

66 Grimod de la Reynière, *Almanach des Gourmands*, 2nd ed., vol. 3 (Paris: Chez Maradan, 1806), 162.

67 Brillat-Savarin, 347.

68 Brillat-Savarin, 34.

69 Grimod de la Reynière, *Almanach des Gourmands*, vol. 6 (Paris: Chez Maradan, 1808), 224.

70 Brillat-Savarin, *Physiology of Taste*, 182.

71 Brillat-Savarin, 15.

72 Michel Foucault, *The History of Sexuality: An Introduction*. (New York: Vintage Books, 1990).

References

Alderman. *The Knife and Fork*. London: H. Hurst, 1849.

Anderson, Benedict. *Imagined Communities: Reflections on the Origin and Spread of Nationalism*. Rev. and Extended ed. London; New York: Verso, 1991.

Assouly, Olivier. *Le Capitalisme esthétique: Essai sur l'industrialisation du goût*. Paris: Editions du Cerf, 2008.

Beauvilliers, Antoine. *L'Art du cuisinier*. Paris: Pilet, 1814.

Berchoux, Joseph. *La Gastronomie*. Paris: Giguet et Michaud, 1805.

Briffault, Eugène Victor. *Paris à table*. Paris: J. Hetzel, 1846. http://catalog.hathitrust.org/Record/006546806.

Brillat-Savarin, Jean Anthelme. *The Physiology of Taste, or, Meditations on Transcendental Gastronomy*. Translated by M.F.K. Fisher. New York: Vintage Books, 2011.

Cadet de Gassicourt, Charles-Louis. *Cours gastronomique, ou, les Diners de Manant-Ville/Ouvrage anecdotique, philosophique et littéraire*. 2nd ed. Paris: Capelle et Renand, 1809.

Carême, Marie-Antoine. *Le Cuisinier parisien, ou, l'Art de la cuisine française au dix-neuvième siècle*. 3rd ed. Vol. 1. Paris: Renouard, 1842.

———. *L'Art de la cuisine française au XIXe siècle. Traité elementaire et pratique*. Vol. II. Paris: Comptoir des Imprimeurs-Unis, 1847.

Chatillon-Plessis. *La vie à table à la fin du XIXe siècle: Théorie, pratique et historique de gastronomie moderne*. Paris: Firmin-Didot, 1894.

Danius, Sara. *The Senses of Modernism: Technology, Perception, and Aesthetics*. Ithaca, NY: Cornell University Press, 2002.

Escoffier, Auguste. *Le Guide culinaire: Aide-mémoire de cuisine pratique*. Paris: Flammarion, 2004.

Foucault, Michel. *The History of Sexuality: An Introduction*. New York: Vintage Books, 1990.

Gouffé, Jules. *Le livre de cuisine*. Paris: Hachette, 1867.

Grimod de la Reynière, Alexandre-Balthazar-Laurent. *Manuel des Amphitryons*. Paris: Capelle et Renard, 1808a.

———. *Almanach des Gourmands: Servant de guide dans les moyens de faire excellente chère*. 2nd ed. Vol. 2. Paris: Chez Maradan, 1805a.

————. *Almanach des Gourmands: Servant de guide dans les moyens de faire excellente chère*. 2nd ed. Revue, corrigée et augmentée, Vol. 2. Paris: Chez Maradan, 1805b.

————. *Almanach des Gourmands: Servant de guide dans les moyens de faire excellente chère*. 2nd ed. Vol. 3. Paris: Chez Maradan, 1806.

————. *Almanach des Gourmands: Servant de guide dans les moyens de faire excellente chère*. Vol. 6. Paris: Chez Maradan, 1808b.

————. *Almanach des Gourmands: Servant de guide dans les moyens de faire excellente chère*. 2nd ed. Vol. 2. Paris: Maradan, 1805c.

Hoffmann, Viktoria von. *From Gluttony to Enlightenment: The World of Taste in Early Modern Europe*. Urbana, IL: University of Illinois Press, 2016.

Humelbergius Secundus, Dick. *Apician Morsels: Or, Tales of the Table, Kitchen, and Larder*. London: Whittaker, 1829.

La Chapelle, Vincent. *Le Cuisinier moderne*. Edited by Luzarches D. Morcrette. 2nd ed. Vol. 5. 5 vols. La Haye, 1984.

————. *Le Cuisinier moderne, qui aprend à donner toutes sortes de repas*. Vol. 1. 4 vols. La Haye: De Groot, 1735.

————. *The Modern Cook*. London: Nicolas Prevost, 1733.

La Varenne, François Pierre. *Le Cuisinier françois, Enseignant la manière de bien apprester et assaisonner toutes sortes de viandes….* Paris: Chez Pierre David, 1651.

Laudan, Rachel. "A Kind of Chemistry." *Petits Propos Culinaires* 62 (1999): 8–22.

Lipovetsky, Gilles, and Jean Serroy. *L'Esthétisation du monde: Vivre à l'âge du capitalisme artiste*. Paris: Gallimard, 2013.

L.S.R. *L'Art de bien traiter*. Lyon: Marchand, 1693.

Marin, François. *Les Dons de Comus, ou, l'Art de la cuisine, réduit en pratique*. 3rd ed. Vol. 1. Paris: Chez Pissot, 1758.

Marx, Karl. *Capital: A Critique of Political Economy*. Translated by Ben Fowkes. Vol. 1. London: Penguin, 1976.

————. *Karl Marx Early Writings*. Translated by T.B. Bottomore. New York; Toronto; London: McGraw-Hill, 1964.

Massialot, François. *Le Cuisinier roïal et bourgeois*. New and Expanded. Paris: Claude Prudhomme, 1705.

Massumi, Brian. "The Autonomy of Affect." *Cultural Critique* Autumn, no. 31 (1995): 83. https://doi.org/10.2307/1354446.

Menon. *La Cuisinière bourgeoise, suivie de l'Office a l'usage de tous ceux qui se mêlent de dépenses de maisons*. Bruxelles: Foppens, 1753.

————. *La Science du maitre d'hôtel cuisinier, avec des observations sur la connaissance et proprietés des alimens*. Paris: Compagnie des Libraires Associés, 1768.

Nebrija, Antonio de. "Prólogo." In *Gramática de la lengua castellana*. Salamanca, 1492.

Neirinck, Edmond, and Jean-Pierre Poulain. *Historia de la cocina y de los cocineros: Técnicas culinarias y prácticas de mesa en Francia, de la Edad Media a nuestros días*. Barcelona: Editorial Zendrera Zariquiey, 2001.

Novero, Cecilia. *Antidiets of the Avant-Garde: From Futurist Cooking to Eat Art*. Minneapolis: University of Minnesota Press, 2010.

Ritzer, George. "The Weberian Theory of Rationalization and the McDonaldization of Contemporary Society." In *Illuminating Social Life: Classical and Contemporary Theory Revisited*. Edited by Peter Kivisto, 5th ed. Thousand Oaks, CA: Pine Forge Press, 2011.

Spang, Rebecca L. *The Invention of the Restaurant: Paris and Modern Gastronomic Culture*. Cambridge, MA; London: Harvard University Press, 2001.

Spary, E.C. *Eating the Enlightenment: Food and the Sciences in Paris, 1670–1760*. Chicago: University of Chicago Press, 2014.

Takats, Sean. *The Expert Cook in Enlightenment France*. Baltimore: Johns Hopkins University Press, 2011.

Thackeray, William Makepeace. *The Book of Snobs*. London: Punch Office, 1848.

Union philanthropique culinaire et d'alimentation. *La Cuisine française et etrangère* 1, no. 1 (August 15, 1891).

Union universelle pour le progrès de l'art culinaire. *La Science culinaire*. Paris, 1878–1888.

4
RACIALIZING TASTE

Spices, and the affective experience of taste that they provide and represent, were the scapegoat whose sacrifice founded modern gastronomic subjectivity. Gastronomy bargained away the embodied and affective aspects of the sense of taste that were incompatible with modern epistemology and aesthetics. Highly affective foods and seasonings, like spices and hot peppers, exceeded the parameters of how gastronomic writers defined modern taste. The persistent aromas of spices and the sensation of heat visibly affect eaters in ways that put into question the rational gastronomic ideal of an eater that objectively judges the foods without affecting their taste or being affected by it. The will to control the affectiveness of taste helped to create the idea of a distinct and superior white European race. Once expelled from gastronomy, the use of spices and the affective experience of taste were turned into a marker of the racialized others of modernity. Europeans feared losing the privileges that imperial colonialism afforded them and wanted to see them as natural. They wanted to believe in race as a biological given, but there was a continued anxiety regarding the lack of fixity of identity. Ann Laura Stoler has argued that what sustained European racial membership in the colonies was "a middle-class morality, nationalist sentiments, bourgeois sensibilities, normalized sexuality, and a carefully circumscribed 'milieu' in school and home."[1] I would expand her point to argue that a desensualized and bureaucratized sense of taste was part of the foundation of European racial membership in Europe and helped sustain it in the colonies.

The dramatic reduction of the use of spices in modern cooking made them ever more present in gastronomic texts. Gastronomic discourse regarding spices and affective taste has four interconnected aspects: disavowal, containment, racialization and surveillance. The vociferous disavowal of spices, even as actual spices were never given up fully, paved the way for the racialization of spices and affective taste as markers of so-called inferior races. Modern gastronomic subjects were

DOI: 10.4324/9781003331834-5

compensated for their renunciation of affective taste by many new power-infused pleasures. The most visible pleasure was the opportunity to assert and perform their self-ascribed racial superiority when eating according to gastronomic rules, particularly in the public space of the restaurant. The disavowal and racialization of affective taste also afforded the pleasure of talking about affective taste cultures with disdain, and the enjoyment of affective taste now transformed into an exciting transgression. Gastronomic writing naturalized and policed the global sensory regime that it defined by producing and circulating narratives of the taste cultures of others in ways that confirmed their racist classification. The gastronomic obsession with "authentic" food cultures owes much to the modern fear of "hybridization," which has the potential of tearing down the modern/colonial order that was predicated on essentialist understandings of difference.

The modern notion of taste was in many ways the result of the active engagement of gastronomy with colonial discourse, which led to its racialization. Gastronomy constructed its notion of taste as a bodily discipline for modern subjects. This discipline from the very beginning served to distinguish the modern from the traditional, the civilized from the uncivilized and the colonizers from the colonized. Gastronomy created a global sensory regime by establishing and policing a racialized understanding of the diversity of human ways of experiencing taste. The modern desensualized approach to taste constructed in gastronomic writing was posited as characteristic of the most advanced social classes and nations, thus casting other approaches to taste as backward. The construction of modern taste was predicated on the racialized hierarchization of taste differences.

The history of gastronomy has been generally understood as a part of the civilizing process. This process was succinctly summarized by Norbert Elias as "everything in which Western society of the last two or three centuries believes itself superior to earlier societies or 'more primitive' contemporary ones."[2] But, as Walter Mignolo has argued, there is a complicity between the "civilizing mission" articulated in colonial discourse and the "civilizing process" articulated as an object of study of the human sciences.[3] In the study of the civilizing process, civilization becomes the exclusive product and property of the West, and the lack of the rest. This Eurocentric understanding of the global history of human civilization has been challenged by many scholars, like Enrique Dussel who explains the rise of Europe as the result of the subalternization of other peoples and cultures and the appropriation of their knowledge, resources and labor.[4] Gastronomy, as a part of the civilizing process, played a role in the articulation of the civilizing mission. Peoples whose taste cultures differed from the one prescribed by gastronomy were marked as inferior biologically, morally and intellectually, and therefore available for colonization. It is well known that gastronomy is among the many cultural consumption practices that serve to legitimize and naturalize social differences. Pierre Bourdieu has explained how the regime of aesthetic taste serves to organize and police the social order inside modern societies.[5] While Elias and Bourdieu are correct in approaching table manners and aesthetic taste as tools that define and police the boundaries of Western bourgeois subjectivity, they did not fully engage with the global colonial and imperial dimension of that subjectivity. The constitution and conditions of possibility of modern subjectivity, global capitalism and gastronomic practice depend on the otherization and subordination of the rest of the world, materially and epistemologically. Gastronomy has been an instrument for the subordination of different ways of sensing and knowing.

Simon Gikandi has explained how, in the eighteenth century, British discourses on aesthetic taste attempted to use culture to conceal the connection between modern subjectivity and the political economy of slavery.[6] He argues that for European high culture to be elevated to the position it has occupied since the early modern period, it needed to debase the black other as the counterpoint to refinement.[7] Africans were "confined to a sensory order mired in dirt, mud, odor, and simple bad taste."[8] But the debasement of sensory orders that had not been submitted to the modern discipline of taste was not limited to the case of enslaved Africans. In the nineteenth century, Hegelian philosophy and narratives of civilization authorized colonialism and naturalized the subjugation of indigenous populations.[9] Colonialism and genocide were seen as perhaps unfortunate but nevertheless inevitable historical developments in the march of progress. In line with the Hegelian philosophy of history, the gustatory taste of modern Europeans, framed as superior, was taken as corresponding to superior bodies and minds that were naturally destined to rule others. The modern Western representation of other sensory orders as degraded was applied to all the peoples subjected to European domination. This includes indigenous peoples, peoples of the "Orient," and "Third World" peoples. Inside of modern Western societies, women, children and the poor were also conceptualized as having an inferior sensorium in need of regulation. Modern Europeans assigned themselves a restrained and non-affective approach to the experience of taste, which they characterized as refined and civilized. In contrast, everybody else was presumed to have bad taste, which was characterized as primitive in the case of Africans and indigenous peoples, and as overindulgent and morally suspect in the case of Asians.

By inserting gustatory taste and culinary cultures into the larger modern narratives of progress and civilization, gastronomy provided a sensorial dimension to the global colonial capitalist order. Gastronomy was deeply indebted to this order for the abundance and variety of foodstuffs and culinary knowledges that made modern consumption possible. These include commodities like sugar, tea, tobacco and coffee, whose rituals of consumption epitomize modernity. Gastronomy also owes the modicum of intellectual and aesthetic legitimacy that it gained for the sense of taste in Europe to the debasement of the cultures of taste of the peoples on which it depended. The construction of modern taste and its commitment to controlling affect was not an innocent expression of a difference in thought and taste. It was part of the process of giving the global imperial order a semblance of righteousness. The modern Western sensorium was profoundly marked by the modern construction of race, which served to legitimize the global modern colonial and capitalist order as biologically determined. Colonial categories of race defined not only aesthetic taste and table manners but culinary practices and the experience of gustatory taste itself.

The Disavowal of Spices

Spices are at the core of the racialized global sensory regime instituted by gastronomy. The founding gesture of modern gastronomy was the rejection of the abundant spice use that characterized medieval European court cuisine. People are often amazed to learn that medieval European cuisines used spices abundantly. However, as noticeable as the reduction in the use of spices was, the most important change was not a change in quantity. The most dramatic change relates to the development of a new interpretive framework that racialized spices and the affective experience of taste. The rejection of spices was stridently voiced in gastronomic writing, but the actual reduction in the use of spices in European cuisines was a very slow and never completed process. Modern Europeans were faster to congratulate themselves for giving up spices than they were to actually give them up. The fervent desire to distance themselves from spice-using peoples is more important than the actual reduction in spice use. This betrays that something more than an innocent taste preference was at stake. The disavowal of spices in Western culinary cultures was not a mere cultural preference, but a powerful political act. The very identity of Europe and its self-perception as a superior civilization depended on the much-flaunted shift away from spiced and otherwise affective foods.

What is a spice? The word in general refers to plant substances used to flavor food, although for a long time it referred exclusively to those that came to Europe through the Indian spice trade, like black pepper, cloves, cinnamon and nutmeg. However, spices have never been a stable category corresponding to a clearly delimited material reality. What is or is not a spice has been a shifting discursive construction. The German word *genussmittel*, which refers to articles of pleasure, includes spices together with condiments, stimulants, intoxicants and narcotics. It refers to affective substances that are consumed for pleasure rather than necessity.[10] The affective power of spices led to their use as medicine in Ayurveda and in Hippocratic humorology, but in modern gastronomy they were conceptualized as similar to alluring but dangerous recreational drugs that could be destructive and should be handled with care.

I argue that spices in gastronomic discourse represent the affective excess of taste that gastronomy was committed to controlling. The affective excess could come from spices and other strong seasonings but also from all foods and ways of eating that affect the body in ways that challenge the codes of bourgeois politeness. Foods that make eaters produce sounds and smells, and foods that stain or make eaters sweat, all fall into the category of highly affective foods that have no space in the gastronomic understanding of taste as disembodied. Affective foods did not disappear from the gastronomic landscape, however. Instead, they became racialized and enjoyed as the somehow dangerous food of inferior others. Gastronomy changed the meaning of spices from the almost unattainable luxury of faraway high civilizations that they were in the Middle Ages, to the alluring but dangerous sense of taste of those that modernity marginalized and exploited.

Scholars have strived to explain the banishment of spices in Europe in a variety of ways, but they tend to echo the gastronomic belief in the superiority and inevitability of the modern aversion to spices, as if it were the evolutionary destiny of humanity to shun them. Medieval Europeans were the first to be otherized by the modern ideology of spices. The avoidance of spices is so central to the modern Western sense of self that many people find it hard to accept that European medieval court kitchens employed spices liberally. Christian Boudan, in his book about the geopolitics of taste, characterized Medieval Europe's taste for spice as "fragile," dismissing it as a temporary detour before they went back to their own flavors and techniques, which were preferred since antiquity.[11] Boudan, repeating the classicist wishing away of the Middle Ages that is part of the narrative of modernity, posits the rejection of spices as Europe reconnecting with its true self.

Fernand Braudel, usually considered a pioneer in scholarly food studies, legitimized many aspects of the modern ideology regarding spices. Braudel stated that spices and all seasonings are used by those condemned to poor and monotonous diets.[12] He implies that people who have an abundant and varied food supply have no need for spices and seasonings. Braudel referred to the use of spices in medieval Europe as a "spice orgy" and stated that "the badly preserved and not always tender meat cried out for the seasoning of strong peppers and spicy sauces, which disguised its poor quality."[13] For Braudel, spices are used only if there is something lacking about the food and therefore they are unnecessary when food quality is good. The use of the words "orgy" and "disguise" reveals the moralism behind the modern avoidance of spices as representatives of unrestrained pleasure.

Another historian who uncritically took the modern disembodied and affectless approach to taste as a natural sign of modern superiority was Reay Tannahill, who declared, "A medieval banquet can have been restful to none of the senses."[14] The perspective of Tannahill's assessment of medieval banquets implies that a banquet should be restful for, rather than engage, the senses. She described the festive and busy environment of the banquets as chaotic. Tannahill included spices in the list of the unpleasant sensory experiences endured at banquets: "But despite the noise, the bustle, the spicy aromas of the food and the no less spicy aroma of unwashed humanity, the courtesy books still found it necessary to make mention of digestive wind."[15] For Tannahill, the smell of spices is akin to bodily odors, and therefore undesirable. The affective aroma of spices is a reminder of our embodied existence, which is something that modern rationalism attempted to suppress.

Other scholars have presented more nuanced explanations for the reduced use of spices in Europe. Stephen Mennell disputed the notion that medieval people were forced to use spices to disguise spoiled meat.[16] He also clarified that the shift in taste in the modern period was more gradual than usually acknowledged.[17] More recently, Paul Freedman has explained that both the use and the eventual abandonment of spices in European cooking have multiple explanations.[18] After discussing the cultural associations that medieval Europeans attached to spices, Freedman still concluded that the decline in the use of spices was gradual but ultimately "What took place was a seismic shift in taste."[19] I argue that such a seismic change cannot be fully explained without taking into account the new cultural associations that modernity attached to spices and the strong ideological investment that the modern West placed on their disavowal.

In the Middle Ages, spices signified luxury and medical know-how, and they had religious associations with paradise.[20] But as modernity developed, new powerful meanings were given to spices through their racialization, which made spices enjoyable in Europe in totally new ways. The dramatic change in the meanings of spices mirrors the equally dramatic change in the geopolitical position of Europe. In the Middle Ages, Europe looked up to the East for all kinds of cultural sophistication but, as Europe acquired the position of centrality in the emerging modern global order, they began to look down on the East and everybody else. While a few scholars have argued that the abandonment of the use of spices was related to the wish to distance European culture from its debt to the Arabic world,[21] others have dismissed that notion.[22] The fact that scholars continue to debate this matter is the result of the unwillingness or inability to give up the modern Western fantasy in which European culture does not owe anything to others, much less to those cast as inferior in modern Western thought.

As Freedman explains, the use of spices in Europe predates Islam and Arab expansion.[23] However, what matters is not just the presence of spices but how and why they were used. In the Middle Ages, Europe used spices in emulation of the sophistication of the East. However, as Europeans solidified their global hegemony, spices for them became a symbol of the excesses that they thought ushered in the demise of distant kingdoms and of their own previously all-powerful aristocracies. Spices became associated with uncontrolled pleasure and decadence. Conversely, abstention from spices was taken as a demonstrable sign of the supposedly superior European taste, bodies and reason. The abstention from spices distanced European culture from the world that it constructed and shunned as the voluptuous Orient. More generally, it placed modern Europeans on a pedestal above all other peoples. Spices served to mark all kinds of others. In the modern discipline of taste, enjoyment of the affective power of food and spices became an observable mark of inferiority which made people a target of local and global disciplinary designs.

The history of the transformation of the meanings associated with the use of spices in Europe is long. The extant cookbooks of the Middle Ages document the use of spices in court cookery across Europe. Sometimes recipes call for a specific spice, but more often than not spices in these cookbooks are treated as indistinct and interchangeable. There doesn't seem to be a sophisticated underlying culinary theory of spice use capable of distinguishing specific uses for different spices and combinations. In the fifteenth century, Bartolomeo Platina (1421–1481) defined spices as imports from abroad.[24] For Platina, spices are somehow dangerous if not used properly. Even before the dramatic changes in the appreciation of spices, they were already defined first and foremost as foreign and somewhat dangerous. But in the Middle Ages, they were luxurious, expensive and sought after since medicine and upscale cuisine were unthinkable without them. This situation changed as spices became more readily available. Spice consumption grew and spread down the social scale, peaking in the late eighteenth century.[25] After that, spice consumption decreased noticeably without ever fully disappearing. The wider availability of spices and the collapse of humoral dietetics contributed to the waning of spice use in European cooking. However, these explanations cannot account for the intensity of the claim of having abandoned spices. The racialization of spices can better explain the ambivalent attitude toward spices that is the trademark of modern Western taste cultures. Gastronomy gave up affective taste for the sake of epistemic validity in the eyes of modern thought and as a way of asserting the racial superiority that the colonial order granted to Europeans. Spices became the main repository of affective taste and continued to be alluring as purveyors of the gustatory pleasures forbidden for the sake of modernity.

Moralist condemnation of the pleasures of taste continued to be the norm throughout the modern period, whether in religious or secular guise. The reason why it was important to pay at least lip service to the need to reduce the use of spices was because of the hostile attitude toward gustatory pleasure of many prominent thinkers. The opposition to gustatory pleasure in a sector of Enlightenment intellectuals targeted all kinds of cuisine that aimed to flatter sensuality beyond the satisfaction of the basic need to eat. Louis de Jaucourt (1704–1779) in his entry on "Cuisine" in the *Encyclopédie* condemned all cuisine as a corrupting luxury.[26] The *Nouveau dictionnaire* of 1776 proudly declared that in France, spices had been proscribed and that they demanded that the seasoning of food be "almost imperceptible."[27] Voltaire (1694–1778), for his part, wrote an article on aesthetic taste in which he defined physical bad taste as the one that is flattered only by spicy and unusual seasonings.[28] Highly seasoned foods were condemned as damaging the palate and making it need increasingly strong seasonings to be satisfied. As E.C. Spary explains, the deterioration of physiological taste was seen as a sign of moral shortcoming and weak reason. Medical authors attacked luxurious eating as a source of political, moral and physical degeneracy.[29] The new moralism against sensory pleasure changed the stakes of indulgence. Rather than going to hell for gluttony, those who indulged in tasty foods now were perceived as politically, morally and physically inferior. In light of this bleak scenario for the cultivation of gustatory pleasure, cooks and gastronomers strived to come up with a cuisine and a notion of taste that would allow some gustatory pleasure while appeasing the opposition to it.

Disavowing spices was an important part of the gastronomic compromise that secured a modicum of legitimacy for the pursuit of gustatory pleasure in Europe, which is discussed in Chapter 2. Renouncing spices allowed gastronomy to comply with the moderation requirement while developing a new style of cooking that would flatter the sense of taste without using the ingredients that became taboo. Spices were bargained out in the modern struggle against the resistance to gustatory pleasure in Europe. The disavowal of spices led cooks to develop a cuisine that was noticeably different, depending on butter and stock rather than on spices to enhance the flavor of food. Overcoming the resistance to the acceptance of the epistemic and aesthetic qualities of cuisine could have been achieved in many other different ways without renouncing the use of spices and affective taste, but this was the way in which it was done. Cooks and gastronomers took on the task of advocating against the use of spices. However, more than eliminating spices, cooks and gastronomers developed new rules for enjoying them in a controlled way that could be considered rational instead of voluptuous.

In the early modern period, cookbooks began the process of limiting the use of spices. *The French Cook*, published by François Pierre de La Varenne in 1651, is considered the first modern cookbook. La Varenne drastically reduced the use of sugar, saffron and spices that was characteristic of medieval cookbooks.[30] However, other cookbooks published in subsequent decades by Nicolas de Bonnefons,[31] Pierre de Lune,[32] L.S.R.[33] and François Massialot[34] continued to call for spices in many of their recipes. In the eighteenth century, Menon[35] and Vincent La Chapelle continued to use spices even as they advocated for and developed a new kind of cooking. In the second edition to *The Modern Cook* (1742), La Chapelle defended himself against accusations that he uses too many spices with the argument that he presupposed the discretion of the cook.[36] Statements disavowing spices became a requirement to qualify as a modern cook, but in culinary practice spices were more marginalized than excluded.

In the nineteenth century, cooks and gastronomers continued to address spices in their writing. The publication of the poem *La Gastronomie* in 1801 introduced the modern discipline of taste.[37] A few years later, the poem *L'Antigastronomie* (1804) presented the case against gastronomy. The familiar criticism of cuisine as immoral sensuality is once again anchored in the use of spices and seasonings, which the poem characterized as poisons.[38] Alexandre-Balthazar-Laurent Grimod de la Reynière (1758–1837), one of the founders of gastronomic writing, published an article about spices as appetite stimulants. After a review of the spices and herbs commonly used in Paris, he concludes that local herbs should be preferred because they are less dangerous than the exotic spices that come from hot climates and which can be real poisons for Parisians if not used judiciously.[39] Part of what made spices dangerous was their foreign origin. In general, gastronomic writers tried to balance their attraction toward spices with the need to present themselves as rational eaters that would not let themselves be possessed by the affective power of spices. Spary has argued that in the *Cours gastronomique* (1809), Charles-Louis Cadet de Gassicourt demonstrated that scientific, literary and historical erudition was at the base of gastronomy.[40] But Gassicourt's book also illustrates how the validation of gastronomy as rational and scientific was built on the marginalization of spices in French cooking.

The lessons given in the *Cours gastronomique* pose spices as representative of pungency, which is presented as one of seven simple flavors alongside freshness, sweetness, saltiness, acidity, bitterness and astringency.[41] The word "pungent," or *âcre* in the French language original, refers to substances that have a strong or irritant taste or smell. Spices as representatives of this category of flavors lose all individuality and subtlety and become just irritants. The ability of spices to assertively affect the body is what brings them together in the gastronomic mind. In Gassicourt's *Cours gastronomique*, the professor entertains his students with details about the use of spices in the past.[42] Later in the gastronomic lessons, it is stated that philosophes should choose foods with discernment, and all the characters agree that too many spices are currently used, when they should be treated as auxiliaries only.[43] The arguments presented by Gassicourt bring together the points made by many before him. He grants that spices and affective foods are excessive and dangerous and that their use should be controlled by reason and good taste. Like others before him, Gassicourt talks about current spice use as more restrained than in the past, but still in need of further reduction. The structure set up in this narrative made taste-flattering cuisine intellectually acceptable by disavowing spices as the repository of the irrational excess that had made all taste-flattering types of cuisine objectionable. This structure made all cuisines seem irrational and excessive except for French cuisine because it had expelled, or at least reduced, the use of spices.

The Containment of Spices

The continued engagement with spices in gastronomic writing suggests that they were not disavowed because they were disliked, but because they had acquired too much negative symbolic and ideological baggage. In the quest to transform spice use into something local rather than imported, and rational rather than sensuous, cooks focused on the development of a way of using spices in a regulated way. One way in which spices were controlled was to restrict their use to specific kinds of food like sweets and forcemeats, and to a few times of the year like Christmas. Another way was to limit spice use to an all-purpose spice mix. Cooks and entrepreneurs competed to establish their own spice mix as the definitive best way to use spices. Standardized spice mixes aimed to stop the endless combinatory possibilities that individual spices can provide. Spice mixes known as "four spices," "fine spices" and "Parisian spices" were meant to codify spice use, but there was no real consensus regarding their specific contents and proportions. Even though modern cooks criticized medieval cookbooks for using too many spices and using the same seasonings in all foods, they limited the use of spices to certain kinds of dishes but continued to think of spices as a generic group that can be used interchangeably.

Early modern cooks hoped to establish a spice mix as the only one needed to season all kinds of dishes. Their aim was not to explore the individual sensory qualities and culinary possibilities of each spice or to develop the art of tailoring specific spice combinations to different dishes in the way that cooks in other parts of the world had done. Their aim was to codify a single all-purpose mix to prevent the proliferation of ways of using and enjoying spices. In the seventeenth century, Bonnefons recommended a spice mix that he considered "good for all kinds of seasonings without exception."[44] It contained black pepper, ginger, cloves, nutmeg and cinnamon. In the eighteenth century, La Chapelle instructed readers to prepare veal with both fine spices and fine herbs. The latter were proposed as the local alternative to foreign spices. He combined cinnamon, cloves, nutmeg, mace and coriander to make the fine spices mix, while basil and thyme together became fine herbs. One century later, Jourdain Lecointe presented a slightly more targeted use of spices, by providing recipes for three different mixes.[45] The mixes were meant to be used in ragoûts, main dishes and sausages. The mix for ragoûts contained cinnamon, cloves, nutmeg, ginger, fennel and coriander. This mix is consistent with the general use of spices in medieval cookbooks. The mix for main dishes, which is the one labeled as fine spices, is a combination of two mixes. One mix contains cinnamon, nutmeg, cloves, coriander and bay leaves, and the other one was made by combining dried truffles, morels and other kinds of mushrooms. While the first group is similar to the ragoût mix, the second group looks like an effort to indigenize spice use by the addition of mushrooms. However, neither the effort to customize spice combinations for specific kinds of dishes nor the attempt to combine them with mushrooms became widespread. What became more common was the mix of spices with herbs, as can be seen in the combination Lecointe suggested for sausages. His recipe called for coriander, anise, cloves, basil and sage, which is generally consistent with the way in which sausages are currently seasoned.

In the second half of the nineteenth century, a general all-purpose spice mix became the norm. The name and the contents of the mix continued to vary, but they were usually a combination of many of the spices that were favored in the Middle Ages with the addition of a few herbs. Jules Gouffé (1807–1877), in his household kitchen manual, gave a list of spices and aromatics that cooks should have available at all times. These included white and gray salts, white and black peppers, nutmeg, cloves, thyme, bay leaf, garlic, ordinary and English mustards, cayenne and chili peppers, garlic oil, Orléans, tarragon and chili vinegars, cinnamon, vanilla, groats flour, orange blossom water, white sugar, caramel, and composed spices.[46] This list is notable for its variety and eclecticism, containing flavorings from all over the globe. Many of the spices favored in the Middle Ages continue to be present, and they are called for throughout the book both individually and as the better part of the recipe for composed spices. Gouffé's recipe for composed spices combined a total of 42 grams of nutmeg, cloves, white pepper and cayenne with 24 grams worth of a combination of thyme, marjoram, bay leaf and rosemary.[47] Spices did not disappear in spite of the efforts to banish them, but cooks strived to simplify and restrict their use.

Food entrepreneurs profited from the impulse to limit spice use to a single standardized spice mix. A cookbook published in Paris in 1872 carried an advertisement for a spice mix called "cook's spices," composed by M.D. Honnoré.[48] The advertisement narrative presents the mix as a much-needed help for cooks that cannot find the perfect combination of seasonings, and as a relief for gourmets who fear the ill effects of imperfect seasoning. This mix is said to be the result of extensive research and to be distinguished by a fine taste that does not irritate the digestive organs. This narrative reveals the ambivalence that modern cooks and gastronomers felt towards spices. On the one hand, they realized that spices are powerful seasonings that cooks should be able to use. On the other hand, it seems like the mastery of any sophisticated theory of spice use was beyond what could be expected from French cooks. A single all-purpose mix was seen as the solution to the problem of how to use spices in a modern way.

A slightly more successful attempt to codify an all-purpose spice mix was made by Léon Cieux. The journal of the Universal Union for the Progress of the Culinary Arts praised the "Parisian spices" that Cieux presented at the food exposition that took place in Paris in 1883.[49] Léon Cieux is also credited with the invention of "nutritive capsules" and "bouillon tablets," and as the author of *L'Art culinaire simplifié*. Cieux was a chef and culinary entrepreneur keen on simplifying the art of cooking. The Parisian spices aimed to give modern legitimacy to the same old generic mix of all-purpose spices by presenting it at the exposition as an advancement. The only description given of the mix of spices is that "they are strong and aromatic without anything dominating."[50] The generic spice mix, now labeled as Parisian, acquired a local and scientific façade. The pretension was that branded spices were scientifically developed and could substitute and improve on the way in which other peoples use spices. Parisian spices had a measure of success. Even Auguste Escoffier (1846–1935) made a point of recommending "Cieux spices" in his canonical cookbook *Le Guide culinaire* (1903), which continues to be treated with reverence by French-trained chefs.[51]

While the Parisian spice mix has been almost forgotten, the most successful all-purpose spice mix has no doubt been curry powder. Curry powder is the result of the interactions between British colonizers and Indian cooks.[52] Even though it is a gross simplification of the highly developed art of spice use in South Asian cuisines, curry powder became the main way in which spices are experienced in the modern West. Curry powder helped to perpetuate the lack of knowledge and sophistication about spice use in Europe that was the result of centuries of embattling and disavowing them. Today many people are surprised to find out that curry is not a single spice but a mix of spices. Gastronomic writing and cookbooks flattened out the diversity of spice combinations and spiced dishes in different taste cultures under the name of curry. The continued use of the word "curry" to refer to any dish that contains a sauce in which spices were used homogenized vastly different dishes from many different places. This homogenization is only possible by failing or refusing to recognize that both spice use and its enjoyment require competence in other culinary knowledges and conceptualizations of taste. Without such competence, all cuisines that use spices seem the same. European cuisine never developed a sophisticated culinary use of spices, and the colonialist attitude of gastronomic writers blinded them to the possibility that other peoples had developed sophisticated theories of spice use that they just could not understand or appreciate without some effort.

French gastronomers congratulated themselves for their vast nomenclature of composed dishes but failed to realize that what they called curries were composed dishes of great variety and extensive nomenclature. Rather than seeing their incapacity to understand and enjoy other cuisines as the result of their own lack of knowledge and skill, they reduced them to caricatures. Gastronomers set out to simplify the cuisine of others in ways that they could easily reproduce and consume. However, the huge diversity of dishes that gastronomers attempted to capture under the name of curry invariably resisted the violence of the classification. The author of an article published in the culinary trade press in 1884 showed exasperation at the lack of codification of the recipe for Indian chicken curry.[53] He decried the lack of theoretical clarity that allows the same name to be used for different dishes and the same dish to go by different names. After describing a few different ways in which he had seen chicken curry be prepared, he offered the procedure that he witnessed from an Indian cook as a way of starting a discussion to settle the matter. His main concern is that the cooking profession will never be respected without clearly codified recipes. However, his impulse was not to explore the many different Indian dishes that have been lumped together as chicken curry but rather to just settle for a single one as if only one way were possible or desirable. The same journal published other chicken curry recipes. Another contributor to the journal offered his own curry recipes after establishing his authority in the matter as someone who had lived in India. He gives additional bits of information regarding his experience of curry, as a way of establishing his recipes as the most "authentically exotic."[54] In general, the many different preparations that were presented as curry did not shake their fervent belief in the idea that there could be only one.

In the British context, curry became more thoroughly incorporated into the local culinary repertoire. The British identification with curry is an expression of the peculiar colonial cosmopolitanism of gastronomy. Embracing the food of India, or at least a flattened version of it, was one more way of asserting dominance over a colonial possession. Eating curry was often embraced as a performative act of domination of a sovereign subject over an object. Because the British eat curry as conquerors, their use of spice is not seen as sliding down in the civilizational and racial scales. Instead, the British embracing of spice is seen as cosmopolitanism. The author of *Apician Morsels* (1829) proudly praised curry as a sauce that makes anything taste good and hints that it makes Britain a better food destination than Paris.[55] It could be argued that mastery of curry is the culinary crowning of the British empire. In 1906, the British gastronomic journal *The Table* conducted a contest to choose the best curry recipe. The winner of the contest was a Miss Wakeford, whose recipe used a branded curry powder.[56] This contest is evidence of the sense of ownership that British gastronomers felt over curry. They considered themselves the masters and judges of a preparation that is taken as the essence of Indian cooking. A few issues after the publication of the results of the curry contest, the editor of *The Table* lamented that the curries served in London restaurants were inferior to the ones home cooked in India because lazy cooks did not know that more than curry powder is required.[57] This did not stop the editor, A.B. Marshall, from marketing her own brand of curry powder.

Like French gastronomers, British gastronomic writers oscillated between the anxiety of knowing that there is more to Indian cooking and spice use than they know, and the willingness to settle for a simplified version while pretending to have mastery. The use of spices was exorcised from modern cooking, and it was marginalized as something that could be easily mastered. Easy mastery of spices was accomplished by negating the existence of aesthetic and culinary theories about them elsewhere and reducing their culinary applications to the use of a generic mix.

The success of curry powder as the principal method of using spices in modern cooking allowed for the enjoyment of spices in a controlled way. This allowed modern cooks to seem knowledgeable about spices without opening the gates of endless gustatory exploration that they feared and constructed as incompatible with the rational and desensualized notion of modern taste.

The Racialization of Gustatory Differences

The disavowal and containment of spices is the foundation of the gastronomic racialization of taste. Gastronomic discourse added a sensorial dimension to modern racist thought by turning openly sensual and affective taste into the taste of inferior races. The gastronomic enjoyment of affective foods and substances was possible, but it was framed as an exciting and dangerous transgression. The fate of spices in Europe was different from the fate of other highly affective substances that were known as foreign stimulants. Spices were the target of discursive attacks, but coffee, tea, tobacco and sugar were enthusiastically embraced as refined habits for the wealthy, and as fuel and anesthesia for the working classes. Spices were not incorporated in the same way as other foreign stimulants because they were harder to conceptualize as new and modern, given their presence in Europe for centuries. Furthermore, the presence of spices in Europe was the result of Europe as a region under the influence of other regions, while the other stimulants arrived or became abundant as the result of triumphant European colonial empires. Spices symbolized the weakness of being influenced while other stimulants symbolized the power of commanding production and commerce all over the world. However, even though other stimulant substances were incorporated into modern gastronomic culture with considerably less resistance than spices faced, they were still considered as in need of careful handling. Gastronomic writers often attributed wildly exaggerated effects to the consumption of the stimulant substances that came from the others of modernity. The pleasure afforded by these substances was in large part the result of the perceived dangers of letting the body be affected by the nature and culture of racialized others.

The incorporation of foods from abroad was a source of anxiety in the colonies and in Europe from the Renaissance onward. The early modern body was conceptualized by humoral medicine as fluid and, since food created bodily differences, colonizers were afraid their bodies might change if they ate the foods of the colonized. Colonizers feared that eating like colonized peoples would transform their bodies to make them more obviously similar to the colonized, which would undermine the idea of the natural superiority of Europeans on which colonial enterprises were premised.[58] Foods had the potential to transform the body and overturn tradition and the local social order.[59] This fear was particularly strong with regard to substances like spices, chocolate, coffee and tobacco. All foods from abroad effected a culinary and cultural influence. Marcy Norton, in her study of how the Amerindian practices and meanings associated with the consumption of chocolate and tobacco were transferred to Europe, demonstrated that the "enchanted" character of these substances prevailed under the attempts to make their consumption seem "rational."[60] Other stimulants were also incorporated into the fabric of European societies, but their acceptance depended on the mystification of the substances and their places and cultures of origin. The mystique of substances like coffee and tobacco came not only from their stimulant properties but from their associations with colonized peoples. Consuming colonial products allowed people in Europe to comfortably and safely partake in the exhilaration of the power and dangers of being a colonizer without leaving Europe. Designing rules for their consumption allowed them to enjoy an affective experience, without losing their claims to the gastronomic restraint that guaranteed their membership in civilized society.

In the nineteenth century, when the conceptualization of the body was changing from the fluid humoral body to the fixed body proposed by the modern discourse of race, the fear of the effects of foreign foods did not disappear. For example, E.M. Collingham has shown that among the British in India between 1800 and 1947, food anxieties similar to the ones faced by the Spanish in the Americas centuries earlier were common.[61] Jean Anthelme Brillat-Savarin (1755–1826), in his famous *Physiology of Taste*, discussed very few food items in detail, but coffee and chocolate were among them, together with other foreign foods like sugar and turkeys. The fact that so many of the few foods addressed by Brillat-Savarin have foreign origin and had become widely available as a result of imperial commerce underlines the importance of colonial novelties for gastronomy. For Brillat-Savarin, these foods are marvelous like sugar, which "ruins nothing," or marvelous and dangerous at the same time, like coffee, which in large quantities would make a man imbecilic or die of consumption.[62]

The attraction and fear toward modern colonial stimulants are clearly presented by novelist Honoré de Balzac in his *Treatise on Modern Stimulants*, which was originally published as an appendix to Brillat-Savarin's *Physiology of Taste* in 1825.[63] The effects of tobacco and coffee on the body fascinate and terrify Balzac, who was a prodigious coffee drinker. While a few of the effects that Balzac attributed to modern stimulants are factual and obvious, most of them are fanciful fears backed up by anecdotes presented as scientific evidence. Balzac greatly exaggerates both the dangers and pleasures provided by the stimulants. The intensity of the pleasure is both the cause and the result of the danger. More than once, Balzac praises "Orientals" as being superior to Europeans when it comes to material pleasures. However, partaking in the oversized pleasures of other peoples could destroy not only the body that consumed them but the future generations and the nations themselves. Impotence, political and otherwise, was the main concern of Balzac regarding the consumption of modern stimulants. Kingdoms had fallen and entire classes and nations were weak and degenerated because of them. In modern taste culture, spices and all kinds of foods and substances that provide highly affective gustatory and olfactory enjoyment were racialized as the physically and politically dangerous –yet infinitely alluring – taste of subjugated others.

Gastronomic writers ordered the gustatory and culinary diversity of the world according to racial hierarchies. They trained readers to adjudicate racial identity and rank based on eating patterns and behaviors. They also taught readers how to shape their own relationship to food and taste in order to perform and display the superior status that the modern racial global order granted them. Cooks and gastronomers explicitly identified a superior cuisine with a superior civilization. In a speech delivered at a banquet held by the Universal Union for the Progress of the Culinary Arts in 1883, Joseph Favre stressed the important role of cuisine in the march of progress.[64] In the preface to his *Dictionnaire universel de cuisine*, Favre insisted on this idea and declared: "The most civilized peoples have developed the best cuisine, they say. It would be equally logical to say that it is good cuisine that has made peoples intelligent."[65] Gastronomic discourse put in place a way of understanding culinary differences as reflecting different stages of civilization. For Favre, who was the general secretary of *L'Academie de cuisine*, cuisine is the cause and not just the result of civilization. In his understanding, France's good cuisine is evidence of its civilized status, but it is also its cause. Civilized status thus depended on eating in the way prescribed by modern cooks and gastronomers.

The process of establishing a racialized understanding of differences in taste can be seen in culinary texts as early as the seventeenth century. In 1693, L.S.R. criticized the recipes of his rival La Varenne by drawing a clear line between the culinary present and the past. This line was understood as one separating civilization and good taste on one side, and the lack thereof on the other. In a diatribe against La Varenne's book, L.S.R. lists many of his competitor's dishes and concludes: "... and endless other miserable dishes that would be more readily suffered among Arabs and the Maracajá than in a depurated climate like ours in which neatness, delicacy, and good taste are the object and subject of our most solid diligence."[66] Dishes that L.S.R. considered in bad taste were described as suitable for peoples of other continents, characterizing them as having unrefined taste, not just a different kind. Europe, and more specifically France, becomes the only place that is refined and has good taste. In 1705, Massialot matter-of-factly wrote that in other lands people do not know how to properly cook their foods and therefore prepare them in the most disgusting ways. He declared that "good taste reigns only in Europe," where they know how to best cook the foods from everywhere.[67] Massialot's belief in the superiority of modern French cooking made it seem natural and justified for him that France has access to the alimentary bounty of the world.

Nineteenth-century gastronomic writers delighted in presenting any culinary and taste differences as demarcating the line between inferior and superior races. Differences in the kind of mediation used to bring food to the mouth were interpreted as corresponding to different stages of civilization. Eating with fingers, which provides tactile eating pleasure by avoiding the mediation of utensils, was characterized as primitive. Eugène Victor Briffault concluded his *Paris à table* (1846) with the articulation of a civilizational scale based on the use of eating utensils:

> Savages eat with their nails as much as with their teeth; the first sign of softening in their mores can be seen in the use of fingers to take the food without tearing it. Other peoples, more advanced, use chopsticks and sharp

sticks: there is progress; the two-pronged fork is used in northern Europe; in England, they are armed with steel tridents with ivory handles; in France, we have the four-pronged fork: that's the height of civilization.[68]

Briffault's civilizational scale assigns the highest level of civilization to those who eat with the most elaborate mediation. Because of the gastronomic suppression of embodied enjoyment, the tactile appreciation of food was left out of the modern experience of taste. There are echoes of Briffault's schema in our contemporary media commonsense. In 2012, Oprah Winfrey embarrassed herself in a visit to India by asking if they *still* ate with their hands, the presupposition being that they are expected to shed that practice. The disciplinary power of Briffault's schema lies in its ability to serve as a tool to classify people at a glance. It can be easily deployed to put people in the place assigned to them by colonizing modernity, regardless of other characteristics any given individual might have. The less bodily enjoyment a person shows while eating, the more civilized and modern they are supposed to be, and vice versa.

Abstention from affective taste was a way of asserting superiority in the global modern racialized taste order. Readers of gastronomic texts were expected to react with horror when reading about the ways in which the others of modernity ate. This horror would make them feel proud of themselves and come to terms with the disavowal of affective taste that modernity required from them. A brief note on "Oriental Cuisine," published in the journal *Le Gastronome* in 1830, allows us to appreciate the ambivalence that French gastronomers felt regarding the more affective and embodied ways of enjoying the eating experience that they witnessed in other peoples. The article describes four generic "Asians" at table, with a keen eye for the embodied and affective way in which they enjoy their food. While the description is presented as proof that the French are not the only ones who enjoy the pleasures of the table, at times the narrator cannot hide his repulsion at their enjoyment. While granting that the manipulation of foods with fingers is done with artistry, the author declares that not even the mythical dog Cerberus, who guards the gates of the underworld, would accept the morsel made by combining all the different foods together.[69] Regarding the diners' approach to the plate of pilaf, the author observes: "as if it were not enough to eat it, they examine it, they consider it, they touch it; the hostess presses it between her fingers to judge its level of cooking."[70] The diners enjoy their perfectly cooked rice with their mind and all their senses, contrasting with the narrator who thinks that eating it should suffice. The association of "Oriental" food and manners with the devil is again suggested a few sentences later when curry is described in contradictory terms: "curry, that noble Indian ingredient, that tonic par excellence made of chili pepper, long pepper, black pepper, red pepper, cinnamon, nutmeg and clove, infernal mix of spices that we imitate so badly on our *têtes de veau en tortue*."[71] Spices and affective taste in this description are a kind of enjoyment that the gastronomer recognizes but in which he cannot participate. Curry is called infernal because of the sensation of heat that it produces, but the word "infernal" also echoes the earlier reference to hell in the quoted passage. Affective taste evokes in gastronomers the idea of sin, as it was a secular gastronomic sin to engage in the enjoyment of food with the body and not just with the mind. Description of affective ways of eating was a kind of voyeuristic gastronomic pleasure in which the observer enjoyed watching and seeming knowledgeable about the behavior that he was supposed to censor.

In Gassicourt's *Cours gastronomique* (1809), there is a long discussion of the culinary customs of different countries. A character who has traveled only in Europe lists the most distinctive dishes that he tried in each country, like beefsteak and pudding in England, cheese in Holland and sauerkraut in Germany. Another character, who has traveled all over the world, lists dishes that he has encountered in different countries. He mentions details like that in India many people live only on figs, that Californians frequently eat cactus fruit, and that in polar countries they eat whale fat. For an encore, the character mentions having tasted elephant, dog and hippopotamus. The selection of dishes defines the cuisine of other countries by whichever dish is most different from the French norm. The discussion then moves on to how among Muslim peoples plumpness in a woman is admired, in contrast to the desirability of slimness among the French. After listening to the whole discussion about shockingly different culinary and beauty ideals, a young lady concludes with relief: "Ah! I am happy to be French."[72] The lesson from this section of the *Cours gastronomique* taught readers to understand taste differences as confirming the natural superiority of the French, culinarily and otherwise.

Gastronomic journals routinely published reports about the cuisine of distant places. The sources for these reports included chronicles and letters written by colonizers, missionaries, traders, travelers, diplomats and cooks employed abroad. Whatever the sources, the journals took the reports at face value, as objective knowledge. The selection of details was done in a way that reinforced the line that had been drawn between modern taste exemplified by the French and inferior others. The eating of dogs, cats and rats is frequently referenced, sometimes as the only mention-worthy aspect of the cuisines of Asia. Since the eating of these animals is repulsive to Europeans, Asian cuisines as a whole become repulsive by association.

The journal *Le Gastronome: Journal universel du goût* (1830–1831) occasionally had a column called "Physiology of taste of all peoples." These columns were obviously inspired by Brillat-Savarin's book, but they were short and superficial. The point of these columns was not to introduce other cuisines as full systems of knowledge in their own right. The point was to highlight any and all differences and construct them as signs of inferiority. A Chinese banquet is described as abundant but containing bizarre dishes and horror-causing condiments. The author laments that meats are brought to the table already cut up, not allowing for the art of carving to shine in China. He also dismisses the custom of ending a banquet with farcical historical plays, stating that the drinking songs of the French are better for digestion.[73] In the essay about Mexico, readers are informed that all the foods are covered in chili, which foreigners cannot handle, and that Mexicans drink too much pulque in spite of the detestable taste that it gets from the sheepskins in which it is carried. The essay states that in general Mexican foods are less nutritious and have "less strength and juice than in Europe."[74] The author of the description of the cuisine of the "savages of Brazil" wonders whether they really eat without forks, knives and spoons and explains that they "are not yet advanced enough in the scale of civilization to know about the symmetry of services. Dishes are brought without order at the same time."[75] To further stress the characterization of the Tupinamba people as savages, the author declares that what they love the most is human flesh and then proceeds to joke about it. In a column about a dinner in Persia, the chronicler calls Persians barbarians because they do not love beef and don't eat venison. The author, who identifies as a cook, is dismayed that his profession is not much honored in Persia because specialized shops sell roasted meats and prepared rice dishes. He shares his disgust when seeing everyone eating with fingers and finally advises readers: "Oh, my friends, if you ever consider travelling in Persia, bring with you, aside from provisions for a long trip, spoons, forks, table linens; in short bring your whole house with you, rather than expose yourself to the colic that kept me in bed."[76]

Essays like the above quoted were abundant in gastronomic publications. They flagged any details from other taste cultures that contrasted with French preferences and instructed readers to react to them with disapproval and disgust. Any contingent and local French preference was treated as a universal standard. Differences were thus turned into shortcomings. The reaction from readers was expected to be like the one from the young lady in Gassicourt's *Cours gastronomique*: sigh with relief and congratulate themselves for being civilized. While gastronomic essays about the culinary cultures of distant lands are evidence of curiosity and excitement regarding the expansion of culinary knowledge and of gustatory experience possibilities, whatever little knowledge of other taste cultures was reaching gastronomic writers was processed through the racialized colonial interpretive template. Differences in taste were presented in a way that aroused amazement and that ultimately reinforced the sense of superiority that gastronomy offered to its subjects.

Authenticity and the Surveillance of the Racialized Gustatory Order

The apparent cosmopolitanism of gastronomic writers was not driven by a desire to allow other gustatory cultures to teach them different ways of experiencing taste. It was driven by the modern colonial need to produce and accumulate knowledge about subject others. The production of racialized knowledge about others is a trademark of modern colonial epistemology.[77] It has been a way of asserting dominance by defining others according to the role assigned to them in the global imperial order. Gastronomy approached other taste cultures in the same non-affective way that it approached taste. Contact with other taste cultures was not seen as an intersubjective exchange capable of affecting both sides. Instead, it was approached as a relationship of dominance of a sovereign subject over a subordinate object. In the production of knowledge about other taste cultures, France was the pinnacle from whose perspective all cultures were evaluated and ranked. The effect of the gastronomic accumulation of knowledge was the establishment of a racialized understanding of culinary and gustatory diversity as if it were objective scientific knowledge. In this understanding, modern Western bodies disciplined by gastronomy figure as having the most advanced and refined taste in world history.

Gastronomic writers sought to establish a "universal cuisine," which was a global culinary order in which cuisines other than the French were reduced to a few token ingredients and dishes. Brillat-Savarin explains the geopolitics of universal cuisine:

> It is gastronomy which so studies men and things that everything worth being known is carried from one country to another, so that an intelligent and planned feast is like a summing-up of the whole world, where each part is represented by its envoys.[78]

According to Brillat-Savarin, peoples are known by their contribution to the feast. Peoples with minimal or little representation on the feast presumably have not produced anything worth being known. Their presupposition was that only the French had developed a complete and universally valid system of cuisine into which all others could be inserted. The poor representation of non-European culinary cultures in universal cuisine was the result of the willful ignorance of the gastronomers, but they attributed it to the unworthiness of the other cuisines.

The late nineteenth century saw the publication of many ambitious encyclopedic books on food and cooking, like Gouffé's *Livre de cuisine* (1867), which he considered a complete treatise of cookery. While aspiring to codify both the high-class and household cuisines of his times, he indicates that he only chose foreign dishes that could be produced with locally available ingredients. He adapted a selection of foreign preparations not only to accommodate local ingredient availability but also to fit French culinary techniques and perspectives. He explains that he was careful to always apply "the invariable principles of the cuisine of my country, which has not imposed itself apparently all over the world without powerful reasons."[79] The adaptation of foreign cuisines to local ingredients and taste is not remarkable, since people everywhere have always done it. The difference in the case of Gouffé, which illustrates the gastronomic perspective, is the claim that he is doing the adaptation not because of a difference between local and foreign taste, but because the French culinary principles are objectively superior. He considers that sheer greatness is what explains the presence of French cuisine all over the world. The role of Western imperial powers in this spread is not acknowledged. The scarcity of preparations from other parts of the world in Gouffé's book is supposed to be interpreted not as lack of knowledge on the part of the author but as the result of the unworthiness of what is omitted. Other major encyclopedic works that pretend to be universally informed are Alexandre Dumas' *Le Grand Dictionnaire de cuisine* (1873),[80] Urbain Dubois' *Cuisine de tous les pays* (1872)[81] and Joseph Favre's *Dictionnaire universel de cuisine* (1889).[82] In these books, the culinary knowledge available to the authors was organized in a way that places France at top and center. Foods from all over the world are described and evaluated from a French taste perspective, but these opinions are presented as objective facts.

The dismissal of other cuisines in encyclopedic books that presented themselves as objective and universal is not just a reflection of a simple ethnocentrism. French and Western ethnocentrism, propelled by the hard and soft power of the colonial empires, imposed itself worldwide. Colonized peoples were made to internalize the Eurocentric perspectives according to which their own cultures, gastronomic or otherwise, were useless. The elites of many colonized countries accepted the gastronomic discipline of taste, which helped to define their class as worthy colonial intermediaries. Claiming that French cuisine and modern Western taste were superior to all others was not a trivial act. Gastronomic discourse elaborated this claim in a way that turned taste preferences into bodily inscribed marks of racial identity. In so doing, they naturalized the unequal power relations of the global modern order, making it seem naturally ordained.

The relationship between gastronomy and colonialism was accepted by gastronomic writers without hesitation. This relationship was not universally praised, though. In one of the most extensive rebukes of gastronomy, the 1804 poem *L'Antigastronomie*, one of the reasons to censor gastronomy was that for its sake "a single man devours entire peoples."[83] The author wanted readers to confront that the brutality that sustained colonial production and commerce is what made gastronomic abundance and novelties possible. While the author of *L'Antigastronomie* considered this situation immoral, most gastronomic writers were not troubled by it. In the texts of Grimod de la Reynière and Brillat-Savarin, arguably the most influential gastronomic texts of all, gastronomy is fueled by new foods, many of which came from the colonies and which they praised as signs of modern progress. For gastronomic writers, the progress of gastronomy justified any means. They openly celebrated colonialism as the source of alimentary bounty. An article in the journal *Le Gastronome* in 1830 rejoices at the takeover of Argelia because it would make food plentiful in France.[84] The plenty and variety provided by colonialism are constantly praised as a wonder of modern life. In London, the author of a Preface to Thomas Walker's *Aristology* (1881) is pleased that the whole world supplied them with food of every kind much cheaper than before.[85] While Western colonialism and imperialism materially dispossessed other peoples, gastronomy aided their epistemological dispossession by declaring them as bereft of any valuable culinary knowledge. Colonized peoples were portrayed as providing gastronomy with ingredients supposedly without any cultural component. Furthermore, the food habits of other peoples were seen as causes for their colonization. Brillat-Savarin twice mentions how India has been colonized because vegetarianism makes for a weak people.[86] Brillat-Savarin, immersed in a meat-based culinary culture, saw vegetarianism as an aberration.

Gastronomic writers decided that the most radically different aspects of other cuisines were their essential identity and set out to praise and protect that difference under the rubric of "authenticity." Since the distinctive traits that they decided defined other taste cultures are exactly the same ones that were supposed to make them inferior and uncivilized, defending such traits was also defending the colonial order. Europeans were afraid of becoming like the colonized if they ate their food, but they were also concerned about what would happen if supposedly inferior people adopted European eating patterns. If cuisine made people civilized, other peoples could have a claim to civilized status by adopting French cuisine and using it as an argument for the equality that the modern colonial global order denied them. Since civilized status depended on ways of eating, many expressed themselves against teaching others to eat in ways that could help take them out of the subordinated position assigned to them.

Jean-Jacques Rousseau had warned against the acclimatization of exotic plants because it produced monstrosities. Such monstrosities included natural ones like sterile flowers, as well as social ones like the slave-based plantation system.[87] While Rousseau's objection to acclimatization had anti-colonial connotations, a similar argument was developed by those who defended the racialized colonial order. In 1894, Chatillon-Plessis was opposed to the "acclimatization" of subject peoples into French culture. He opposed the practice of giving French food to the people of "exotic races" that were exhibited at the Jardin d'acclimatation. Chatillon-Plessis did not agree with those who protested the treating of people like animals. What upset him was that visitors would give French sweets to the exhibited children through the fence. He considered that "the noble strangers corrupt their spirit and their stomach under the pretext of becoming civilized."[88] It is telling that Chatillon-Plessis was convinced that the reason why the children rushed to eat the sweets was to become civilized. That is clearly his interpretation and his fear. To appease his fear, he decides that civilized French food did not suit the foreigners and could only possibly harm them: "Corruption, I tell you, future toothaches, decadence of taste and health! Poor cherubs!"[89] The fear of the blurring of the line between civilized and uncivilized peoples is at the heart of the seemingly well-intentioned call to protect those characterized as noble savages who should not be allowed to change and demand the equality that was denied them on the grounds of their difference. Culinary authenticity became an important concept for the preservation of the global modern colonial order. Gastronomy has always tended to see differences in taste as biologically and racially determined. Indeed, one of its most famous dictums is "tell me what you eat, and I shall tell you what you are."[90]

The efforts to bring the culinary diversity of the world into the interpretive template of modern gastronomy was done mostly through writing. The international exhibitions that became increasingly popular in the second half of the nineteenth century provided opportunities to put on display the racialized understanding of culinary difference as a spectacle that called into existence the order that had been devised in writing. The stated purpose of these exhibitions was to showcase the cultural and industrial advancements of different countries. Many world fairs focused on industry and others specialized in food and agriculture while others, called colonial expositions, exhibited peoples from the colonial possessions in the imperial metropolises. Regardless of the main theme, all world fairs presented the whole world organized and interpreted by the Western empires. National cultures were not so much represented as constructed at the world fairs, and they were not formed independently but shaped in relationship to one another in the context of empire.[91] The world fairs offered an opportunity to visually and experientially expose a larger public to the racialized understanding of the global culinary and gustatory order.

The fairs also show the limitations of the universalist ambitions of gastronomy. According to Blanchard Jerrold's *Epicure's Year Book* (1868), the kitchens of the Universal Exhibition of Industry that took place in Paris in 1867 were a failure. France had promised "a meeting of the cooks of all nations" and had announced with fanfare that "the kitchens of the universe would be found in an unbroken circle, each perfect in its specialities, on the Champ de Mars."[92] The fair was planned as a perfectly ordered meeting of all cuisines, but the world did not submit so easily to the classificatory dreams of the French. The *Epicure* was disappointed to find gaps in the circle and to realize that the Chinese, Turkish and Tunisian kitchens were poorly executed shams. He was unsatisfied with how England was represented but was pleased by the American and Russian kitchens. The showcase of French wines impressed the *Epicure* with its "magnificent cellars, in which all the good *crus* of the Gironde were laid and carefully catalogued, as though each bottle were a precious volume at a state library."[93] However, French wines seem to have been the only exhibit that complied with the call to orderly classification. The *Epicure* reports that the representation of French cuisine was subpar and overpriced.[94] He quotes a *cordon bleu* who had predicted that French cuisine would be negatively affected by the "meeting of the nations" at the fair: "[…]When we shall have stewed and boiled, without the least regard to art, in order to feed whole nations of strangers who are making a descent on Paris, not one of us will have hand or palate left, our *cuisine* will be at an end."[95] The *cordon bleu* was frustrated by the task of cooking in large quantity, which did not allow for the refined cooking that he would prefer. But there is also discomfort at the idea of feeding foreign masses, presumably incapable of appreciating their efforts. In any case, at least according to the report of the *Epicure*, the 1867 exhibition fell short of projecting the grandiose vision that France had of itself and its cuisine.

Contrasting with the 1867 Universal Exhibition, the 1883 *Exposition Alimentaire* focused exclusively on cuisine and had only a national scope. The journal *La Science culinaire* (1878–1888), which was the official publication of the Union for the Progress of the Culinary Arts, reported extensively on this event. As the journal of cooks rather than gourmands, *La Science culinaire* awaited with excitement the opening of the exhibition, which they hailed as contributing to the progress of their trade by the divulgation of all culinary knowledge.[96] The food fair was the initiative of businesspeople and, although they presented the event as being in the public interest, it shows a shift toward the commercialization of branded products. The commodification impulse first evidenced on Grimod's *Almanach des Gourmands* (1803–1812) is seen at this fair in full force. The main purpose of the fair was to submit products for judgment. The jury gave a large number of medals in many different categories.[97] Presumably these medals would become powerful advertisement tools. But the fair was relatively humble, compared to the botched universalization dreams of the 1867 exhibition. The food exhibition was premised on the idea that France is the gastronomic nation *par excellence* and, according to Favre, it would once again demonstrate that Paris is the powerful head to which others contribute their part of scientific, artistic and economic knowledge.[98] In this fair, France was content with concentrating efforts on the centralization of French food production around Paris, and it abandoned the failed dream of exhibiting all the cuisines of the world.

The presentation of the cuisines of the different countries of the world at the many different world fairs ultimately was split along the civilized versus uncivilized line that gastronomy had drawn following colonial discourse. Modern countries presented themselves at the universal exhibitions while colonized peoples and their foods were displayed in human zoos and colonial exhibitions. These fairs had the obvious goal of validating colonialism as a civilizing mission. Gastronomy provided the interpretative structure for the alimentary customs of colonized peoples to be witnessed and even experienced by the metropolitan masses as a way of confirming their supposed racial superiority. The continuing fascination of gastronomy with culinary authenticity as radical difference has its roots in the modern colonial fantasies of Western racial superiority.

Notes

1 Ann Laura Stoler, *Race and the Education of Desire: Foucault's History of Sexuality and the Colonial Order of Things* (Durham, NC: Duke University Press, 1995), 105.
2 Norbert Elias, *The Civilizing Process: Sociogenetic and Psychogenetic Investigations*, Revised (Oxford: Blackwell Publishing, 2000), 5.
3 Walter Mignolo, *Local Histories/Global Designs* (Princeton: Princeton University Press, 2000), 304.
4 Enrique Dussel, "Beyond Eurocentrism: The World-System and the Limits of Modernity," in *The Cultures of Globalization, Post-Contemporary* Interventions, eds. Fredric Jameson and Masao Miyoshi (Durham, NC: Duke University Press, 1999), 3–31.
5 Pierre Bourdieu, *Distinction: A Social Critique of the Judgement of Taste*, trans. Richard Nice (Cambridge, MA: Harvard University Press, 1984).
6 Simon Gikandi, *Slavery and the Culture of Taste* (Princeton, NJ: Princeton University Press, 2011), loc. 522 of 8044, Kindle.
7 Gikandi, loc. 4215 of 8044, Kindle.
8 Gikandi, loc. 4588 of 8044, Kindle.
9 Lisa Lowe, *The Intimacies of Four Continents* (Durham, NC: Duke University Press, 2015), 147.
10 Wolfgang Schivelbusch, *Tastes of Paradise: A Social History of Spices, Stimulants, and Intoxicants* (New York: Vintage Books, 1993), xiii.
11 Christian Boudan, *Geopolítica del gusto: La guerra culinaria* (Gijón: Trea, 2008), 87.
12 Fernand Braudel, *Civilization and Capitalism, 15th–18th Century* (New York: Harper & Row, 1982), 220.
13 Braudel, 221.
14 Reay Tannahill, *Food in History* (New York: Crown Trade Paperbacks, 1995), 192.
15 Tannahill, 192–193.
16 Stephen Mennell, *All Manners of Food: Eating and Taste in England and France from the Middle Ages to the Present*, 2nd ed. (Urbana, IL: University of Illinois Press, 1996), 53.
17 Mennell, 63–64.
18 Paul Freedman, *Out of the East: Spices and the Medieval Imagination* (New Haven, CT: Yale University Press, 2008).

19 Freedman, 222.

20 Freedman.

21 T. Sarah Peterson, *Acquired Taste: The French Origins of Modern Cooking* (Ithaca, NY: Cornell University Press, 1994).

22 Bruno Laurioux, *Une histoire culinaire du moyen âge,* Sciences, techniques et civilisations du Moyen Âge à l'aube des lumières 8 (Paris: Champion, 2005), 305–335.

23 *Freedman,* 26.

24 Platina and Mary Ella Milham, *Platina, on Right Pleasure and Good Health: A Critical Edition and Translation of De Honesta Voluptate et Valetudine* (Tempe, AZ: Medieval & Renaissance Texts & Studies, 1998), 181.

25 Jack Turner, *Spice: The History of a Temptation* (New York: Vintage Books, 2005), 299.

26 Jaucourt, "Cuisine," in *Encyclopédie, ou, Dictionnaire raisonné des sciences, des arts et des métiers,* eds. Denis Diderot and D'Alembert, vol. Tome quatrième, Conjonctif-Discussion (Paris: Briasson et al., 1751), 537–539.

27 Denis Diderot, et al. *Nouveau dictionnaire, pour servir de supplément aux dictionnaires des sciences, des arts et des métiers,* vol. II (Paris: Panckoucke, 1776), 664.

28 Voltaire, "Goût," in *Dictionnaire philosophique* (Garnier, 1878), https://fr.wikisource.org/wiki/Dictionnaire_philosophique/Garnier_(1878)/Go%C3%BBt.

29 E.C. Spary, *Eating the Enlightenment: Food and the Sciences in Paris, 1670–1760* (Chicago: University of Chicago Press, 2014), loc. 3818 of 9350, Kindle.

30 T. Sarah Peterson, *Acquired Taste: The French Origins of Modern Cooking* (Ithaca, NY: Cornell University Press, 1994), xiii.

31 Nicolas de Bonnefons, *Les Délices de la Campagne, suite du Jardinier françois,* 2nd ed. (Amsteldan [sic]: Raphael Smith, 1655).

32 Pierre de Lune, *Le Cuisinier* (Paris: P. David, 1656).

33 L.S.R., *L'Art de bien traiter* (Lyon: Marchand, 1693).

34 Massialot, François, *Le Cuisinier roïal et bourgeois,* new and expanded (Paris: Claude Prudhomme, 1705).

35 Menon, *La Science du maitre d'hôtel cuisinier, avec des observations sur la connaissance et propriétés des alimens* (Paris: Compagnie des Libraires Associés, 1768).

36 Vincent La Chapelle, *Le Cuisinier moderne,* 2nd ed. (La Haye: Daniel Morcrette, 1742), iii.

37 Joseph Berchoux, *La Gastronomie* (Giguet et Michaud, 1805).

38 Jean-Baptiste Gouriet, *L'Antigastronomie, ou, l'Homme de ville sortant de table, Poëme En IV Chants....* (Paris: Chez Hubert, 1804), 14.

39 Alexandre-Balthazar-Laurent Grimod de la Reynière, *Almanach des Gourmands,* vol. 5 (Paris: Chez Maradan, 1807), 179–196.

40 E.C. Spary, "Making a Science of Taste: The Revolution, the Learned Life, and the Invention of 'Gastronomie,'" in *Consumers and Luxury: Consumer Culture in Europe 1650–1850 Consumers and Luxury: Consumer Culture in Europe 1650–1850,* eds. Maxine Berg and Helen Clifford (Manchester: Manchester University Press, 1999), 179.

41 Charles-Louis Cadet de Gassicourt, *Cours gastronomique, ou, les Diners de Manant-Ville/Ouvrage anecdotique, philosophique et littéraire,* 2nd ed. (Paris: Capelle et Renand, 1809), 64–65.

42 Gassicourt, 127–132.

43 Gassicourt, 294–295.

44 Bonnefons, 42–43. My translation.

45 Jourdain Lecointe, *Le Cuisinier des Cuisiniers,* 10th ed. (Paris: L Maison, 1844), 69–70.

46 Jules Gouffé, *Le Livre de Cuisine* (Paris: Hachette, 1867), 33–34.

47 Gouffé, 34–35.

48 Urbain Dubois, *Cuisine des tous les pays: Études cosmopolites,* 3rd ed. (Paris: E. Dentu, 1872), 599.

49 Union universelle pour le progrès de l'art culinaire, "A travers l'exposition," *La Science culinaire* 6, no. 112 (May 15, 1883): 1–2.

50 Union universelle, 2.

51 Auguste Escoffier, *Le Guide culinaire: Aide-mémoire de cuisine pratique* (Paris: Flammarion, 2004), 165.

52 The complex interactions that led to the invention of curry happened in a colonial context under clearly uneven power relations. However, it should not be assumed that in this interaction the British had absolute power or that the Indian cooks had no agency. For a discussion of different angles, see E.M. Collingham, *Curry: A Tale of Cooks and Conquerors* (Oxford: Oxford University Press, 2006), Cecilia Leong-Salobir, *Food Culture in Colonial Asia: A Taste of Empire* (Abingdon, Oxon; New York: Routledge, 2011), Uma Narayan, "Eating Cultures: Incorporation, Identity and Indian Food," in *Dislocating Cultures: Identities, Traditions, and Third-World Feminism* (New York: Routledge, 1997), 159–188, and Susan Zlotnick, "Domesticating Imperialism: Curry and Cookbooks in Victorian England," *Frontiers: A Journal of Women Studies* 16, no. 2/3 (1996): 51–68, among many others.

53 Ch. Dietrich, "Du poulet au kari a l'indienne," *L'Art culinaire* 2, no. 1 (1884): 27–28.

54 Julien Siméon, "Encore le kari," *L'Art culinaire* 3 (1885): 234–235.

55 Dick Humelbergius Secundus, *Apician Morsels: or, Tales of the Table, Kitchen, and Larder* (London: Whittaker, 1829), 76.

56 A.B. Marshall, ed., "Best Four Recipes for Curry," *The Table* XLII, no. 1065 (November 3, 1906): 20.

57 Marshall, ed., "Current Topics: Indian Curries" XLII, no. 1070 (December 8, 1906): 76.

58 For a detailed discussion of this in the case of Spanish colonizers, see Rebecca Earle, *The Body of the Conquistador: Food, Race, and the Colonial Experience in Spanish America, 1492–1700* (Cambridge, UK; New York: Cambridge University Press, 2013).

59 Spary, *Eating the Enlightenment*, loc. 329 of 9350, Kindle.

60 Marcy Norton, *Sacred Gifts, Profane, Pleasures: A History of Tobacco and Chocolate in the Atlantic World* (Ithaca, NY: Cornell University Press, 2008).

61 E.M. Collingham, *Imperial Bodies: The Physical Experience of the Raj, c. 1800–1947* (Cambridge, UK: Malden, MA: Polity Press; Blackwell Publishers, 2001).

62 Jean Anthelme Brillat-Savarin, *The Physiology of Taste, or, Meditations on Transcendental Gastronomy*, trans. M.F.K. Fisher (New York: Vintage Books, 2011), 113–116.

63 Honoré de Balzac, *Treatise on Modern Stimulants*, trans. Kassy Hayden (Cambridge, MA: Wakefield Press, 2018), 5.

64 Union universelle, "Le Banquet du 13 août," *La Science culinaire* 5, no. 92 (juillet 1882): 2.

65 Joseph Favre, "L'Auteur au Lecteur," in *Dictionnaire universel de cuisine et d'hygiène alimentaire: Modification de l'homme par l'alimentation*, vol. 1 (Paris: Les Librairies, 1889), x. My translation.

66 L.S.R., *L'Art de bien traiter*, 7. My translation. Maracajás, "margajeats" in the French original, is the Tupi name for a jungle wildcat. This word was used in French texts to refer to indigenous groups of Brazil who were enemies of the groups that were friendly to the French. See Jean de Léry and Janet Whatley, *History of a Voyage to the Land of Brazil, Otherwise Called America* (Berkeley, CA: University of California Press, 2006), 235.

67 Massialot, iii–iv.

68 Eugène Victor Briffault, *Paris à table* (Paris: J. Hetzel, 1846), 184, http://catalog.hathitrust.org/Record/006546806. My translation.

69 Paul Lacroix, ed., "Cuisine orientale," *Le Gastronome : Journal universel du goût* 1, no. 65 (October 31, 1830): 3.

70 Lacroix, 4. My translation.

71 Lacroix, 4. My translation.

72 Gassicourt, 162–173. My translation.

73 Paul Lacroix, "Physiologie du goût chez tous les peuples. Chine," *Le Gastronome: Journal universel du goût* 1, no. 26 (June 10, 1830): 1–3.

74 Paul Lacroix, "Physiologie du goût chez tous les peuples. Mexico," *Le Gastronome: Journal universel du goût* 1, no. 29 (June 20, 1830): 2–3.

75 Paul Lacroix, "Physiologie du goût chez tous les peuples. Sauvages du Brésil," *Le Gastronome: Journal universel du goût* 1, no. 31 (June 27, 1830): 3.
76 Paul Lacroix, ed., "Un festin a Ispahan," *Le Gastronome: Journal universel du goût* 2, no. 128 (June 9, 1831): 6. My translation.
77 For a discussion of this idea see Walter D. Mignolo, "The Geopolitics of Knowledge and the Colonial Difference," *The South Atlantic Quarterly* 101, no. 1 (Winter 2002): 57–96.
78 Brillat-Savarin, 63.
79 Jules Gouffé, *Le Livre de cuisine: Comprenant la cuisine de ménage et la grande cuisine* (Paris: Hachette, 1867), 232. My translation.
80 Alexandre Dumas, *Grand dictionnaire de cuisine* (Paris: A. Lemerre, 1873).
81 Dubois.
82 Joseph Favre, *Dictionnaire universel de cuisine et d'hygiène alimentaire: Modification de l'homme par l'alimentation* (Paris: Les Librairies, 1889).
83 Gouriet, 23–24.
84 Paul Lacroix, ed. "Resultat de la prise d'Alger," *Le Gastronome: Journal universel du goût* 1, no. 36 (July 15, 1830): 1–2.
85 Felix Summerly, preface to *Aristology, Or the Art of Dining*, by Thomas Walker (London: George Bell and Sons, 1881), 8.
86 *Brillat-Savarin*, 77 and 156–157.
87 Alexandra Cook, "Jean-Jacques Rousseau and Exotic Botany," *Eighteenth-Century Life* 26, no. 3 (October 1, 2002): 181–201, https://doi.org/10.1215/00982601-26-3-181.
88 Chatillon-Plessis, *La vie à table à la fin du XIXe siècle: Théorie, pratique et historique de gastronomie moderne* (Paris: Firmin-Didot, 1894), 56. My translation.
89 Chatillon-Plessis, 56. My translation.
90 Brillat-Savarin, 15.
91 Carol A. Breckenridge, "The Aesthetics and Politics of Colonial Collecting: India at World Fairs," *Comparative Studies in Society and History* 31, no. 2 (April 1989): 196–197.
92 Blanchard Jerrold, *The Epicure's Year Book and Table Companion* (London: Bradbury, Evans and Co., 1868), 130–131.
93 Jerrold, 136.
94 Jerrold, 130–139.
95 Jerrold, 139.
96 Societé des cuisiniers français, "L'Exposition alimentaire du Cours La Reine (Champs Elysées)," *L'Art culinaire* 6, no. 109 (au 15 avril 1883): 1.
97 Union universelle, "Exposition alimentaire: Recompenses," *La Science culinaire* 6, no. 114 (au 30 juin 1883): 2–3.
98 Joseph Favre, "L'Exposition alimentaire du Cours La Reine (Champs-Elysées)," *La Science culinaire* 6, no. 112 (May 15, 1883): 1.

References

Balzac, Honoré de. *Treatise on Modern Stimulants*. Translated by Kassy Hayde. Cambridge, MA: Wakefield Press, 2018.

Berchoux, Joseph. *La Gastronomie*. Giguet et Michaud, 1805.

Bonnefons, Nicolas de. *Les Délices de la campagne, suite du Jardinier françois*. 2nd ed. Amsteldan [sic]: Raphael Smith, 1655.

Boudan, Christian. *Geopolítica del gusto: La guerra culinaria*. Gijón: Trea, 2008.

Bourdieu, Pierre. *Distinction: A Social Critique of the Judgement of Taste*. Translated by Richard Nice. Cambridge, MA: Harvard University Press, 1984.

Braudel, Fernand. *Civilization and Capitalism, 15th–18th Century*. New York: Harper & Row, 1982.

Breckenridge, Carol A. "The Aesthetics and Politics of Colonial Collecting: India at World Fairs." *Comparative Studies in Society and History* 31, no. 2 (April 1989): 195–216.

Briffault, Eugène Victor. *Paris à table*. Paris: J. Hetzel, 1846. http://catalog.hathitrust.org/Record/006546806.

Brillat-Savarin, Jean Anthelme. *The Physiology of Taste, or, Meditations on Transcendental Gastronomy*. Translated by M.F.K. Fisher. New York: Vintage Books, 2011.

Cadet de Gassicourt, Charles-Louis. *Cours gastronomique, ou, les Diners de Manant-Ville/Ouvrage anecdotique, philosophique et littéraire*. 2nd ed. Paris: Capelle et Renand, 1809.

Chapelle, Vincent de la. *Le Cuisinier moderne*. 2nd ed. La Haye: Daniel Morcrette, 1742.

Chatillon-Plessis. *La vie à table à la fin du XIXe siècle: Théorie, pratique et historique de gastronomie moderne*. Paris: Firmin-Didot, 1894.

Collingham, E.M. *Imperial Bodies: The Physical Experience of the Raj, c. 1800–1947*. Cambridge, UK; Malden, MA: Polity Press; Blackwell Publishers, 2001.

———. *Curry: A Tale of Cooks and Conquerors*. Oxford: Oxford University Press, 2006.

Cook, Alexandra. "Jean-Jacques Rousseau and Exotic Botany." *Eighteenth-Century Life* 26, no. 3 (October 1, 2002): 181–201. https://doi.org/10.1215/00982601-26-3-181.

Diderot, Denis et al. *Nouveau dictionnaire, pour servir de supplément aux dictionnaires des sciences, des arts et des métiers*. Vol. II. Paris: Panckoucke, 1776.

Dietrich, Ch. "Du poulet au kari a l'indienne." *L'Art culinaire* 2, no. 1 (1884): 27–28.

Dubois, Urbain. *Cuisine des tous les pays: Études cosmopolites*. 3rd ed. Paris: E. Dentu, 1872.

Dumas, Alexandre. *Grand dictionnaire de cuisine*. Paris: A. Lemerre, 1873.

Dussel, Enrique D. "Beyond Eurocentrism: The World-System and the Limits of Modernity." In *The Cultures of Globalization*. Edited by Fredric Jameson and Masao Miyoshi, 3–31. Durham, NC: Duke University Press, 1999.

Earle, Rebecca. *The Body of the Conquistador: Food, Race, and the Colonial Experience in Spanish America, 1492–1700*. Cambridge, UK; New York: Cambridge University Press, 2013.

Elias, Norbert. *The Civilizing Process: Sociogenetic and Psychogenetic Investigations*. Revised. Oxford: Blackwell Publishing, 2000.

Escoffier, Auguste. *Le Guide culinaire: Aide-mémoire de cuisine pratique*. Paris: Flammarion, 2004.

Favre, Joseph. *Dictionnaire universel de cuisine et d'hygiène alimentaire: Modification de l'homme par l'alimentation*. Paris: Les Librairies, 1889a.

———. "L'Auteur au Lecteur." In *Dictionnaire universel de cuisine et d'hygiène alimentaire: Modification de l'homme par l'alimentation*, Vol. 1. Paris: Les Librairies, 1889b.

———. "L'Exposition alimentaire du Cours La Reine (Champs-Elysées)." *La Science culinaire* 6, no. 112 (May 15, 1883): 1.

Freedman, Paul H. *Out of the East: Spices and the Medieval Imagination*. New Haven, CT: Yale University Press, 2008.

Gikandi, Simon. *Slavery and the Culture of Taste*. Princeton, NJ: Princeton University Press, 2011.

Gouffé, Jules. *Le Livre de cuisine: Comprenant la cuisine de ménage et la grande cuisine*. Paris: Hachette, 1867.

Gouriet, Jean-Baptiste. *L'Antigastronomie, ou, l'Homme de ville sortant de table, Poëme En IV Chants....* Paris: Chez Hubert, 1804.

Grimod de La Reynière, Alexandre-Balthazar-Laurent. *Almanach des Gourmands: Servant de guide dans les moyens de faire excellente chère*. Vol. 5. Paris: Chez Maradan, 1807.

Humelbergius Secundus, Dick. *Apician Morsels: Or, Tales of the Table, Kitchen, and Larder*. London: Whittaker, 1829.

Jaucourt. "Cuisine." In *Encyclopédie, ou Dictionnaire raisonné des sciences, des arts et des métiers*. Edited by Denis Diderot and D'Alembert, Tome quatrième, Conjonctif-Discussion: 537–539. Paris: Briasson, et al., 1751.

Jerrold, Blanchard. *The Epicure's Year Book and Table Companion*. London: Bradbury, Evans and Co., 1868.

Lacroix, Paul. "Physiologie du goût chez tous les peuples. Chine." *Le Gastronome: Journal universel du goût* 1, no. 26 (June 10, 1830a): 1–3.

———. "Physiologie du goût chez tous les peuples. Mexico." *Le Gastronome: Journal universel du goût* 1, no. 29 (June 20, 1830b): 2–3.

———. "Physiologie du goût chez tous les peuples. Sauvages du Brésil." *Le Gastronome: Journal universel du goût* 1, no. 31 (June 27, 1830c): 2–3.

Lacroix, Paul, ed. "Un festin a Ispahan." *Le Gastronome: Journal universel du goût* 2, no. 128 (June 9, 1831): 5–6.

———. "Resultat de la prise d'Alger." *Le Gastronome: Journal universel du goût* 1, no. 36 (July 15, 1830d): 1–2.

———. "Cuisine orientale." *Le Gastronome: Journal universel du goût* 1, no. 65 (October 31, 1830e): 3–4.

Laurioux, Bruno. *Une histoire culinaire du moyen âge*. Paris: Champion, 2005.

Lecointe, Jourdan. *Le Cuisinier des Cuisiniers*. 10th ed. Paris: L Maison, 1844.

Leong-Salobir, Cecilia. *Food Culture in Colonial Asia: A Taste of Empire* Abingdon, Oxon; New York: Routledge, 2011.

Léry, Jean de, and Janet Whatley. *History of a Voyage to the Land of Brazil, Otherwise Called America*. Berkeley, CA: University of California Press, 2006.

Lowe, Lisa. *The Intimacies of Four Continents*. Durham, NC: Duke University Press, 2015.

L.S.R. *L'Art de bien traiter*. Lyon: Marchand, 1693.

Lune, Pierre de. *Le cuisinier*. Paris: P. David, 1656.

Marshall, A.B. ed. "Best Four Recipes for Curry." *The Table* XLII, no. 1065 (November 3, 1906a): 20.

———. "Current Topics: Indian Curries" XLII, no. 1070 (December 8, 1906b): 76.

Massialot, François. *Le Cuisinier roïal et bourgeois*. New and Expanded. Paris: Claude Prudhomme, 1705.

Mennell, Stephen. *All Manners of Food: Eating and Taste in England and France from the Middle Ages to the Present*. 2nd ed. Urbana, IL: University of Illinois Press, 1996.

Menon. *La Science du maitre d'hôtel cuisinier, avec des observations sur la connaissance et proprietés des alimens*. Paris: Compagnie des Libraires Associés, 1768.

Mignolo, Walter D. *Local Histories/Global Designs*. Princeton, NJ: Princeton University Press, 2000.

———. "The Geopolitics of Knowledge and the Colonial Difference." *The South Atlantic Quarterly* 101, no. 1 (Winter 2002): 57–96.

Narayan, Uma. "Eating Cultures: Incorporation, Identity and Indian Food." In *Dislocating Cultures: Identities, Traditions, and Third-World Feminism*, 159–188. New York: Routledge, 1997.

Norton, Marcy. *Sacred Gifts, Profane, Pleasures: A History of Tobacco and Chocolate in the Atlantic World*. Ithaca, NY: Cornell University Press, 2008.

Peterson, T. Sarah. *Acquired Taste: The French Origins of Modern Cooking*. Ithaca, NY: Cornell University Press, 1994.

Platina, and Mary Ella Milham. *Platina, on Right Pleasure and Good Health: A Critical Edition and Translation of De Honesta Voluptate et Valetudine*. Tempe, AZ: Medieval & Renaissance Texts & Studies, 1998.

Schivelbusch, Wolfgang. *Tastes of Paradise: A Social History of Spices, Stimulants, and Intoxicants*. New York: Vintage Books, 1993.

Siméon, Julien. "Encore le kari." *L'Art culinaire* 3 (1885): 234–235.

Societé des cuisiniers français. "L'Exposition alimentaire du Cours La Reine (Champs Elysées)." *L'Art culinaire* 6, no. 109 (au 15 avril 1883): 1.

Spary, E.C. *Eating the Enlightenment: Food and the Sciences in Paris, 1670–1760*. Chicago: University of Chicago Press, 2014.

———. "Making a Science of Taste: The Revolution, the Learned Life, and the Invention of 'Gastronomie.'" In *Consumers and Luxury: Consumer Culture in Europe 1650–1850*. Edited by Maxine Berg and Helen Clifford, 170–182. Manchester: Manchester University Press, 1999.

Stoler, Ann Laura. *Race and the Education of Desire: Foucault's History of Sexuality and the Colonial Order of Things*. Durham, NC: Duke University Press, 1995.

Summerly, Felix. Preface to *Aristology, Or the Art of Dining*. Edited by Thomas Walker, v–viii. London: George Bell and Sons, 1881.

Tannahill, Reay. *Food in History*. New York: Crown Trade Paperbacks, 1995.

Turner, Jack. *Spice: The History of a Temptation*. New York: Vintage Books, 2005.

Union universelle pour le progrès de l'art culinaire. "Le Banquet du 13 août." *La Science culinaire* 5, no. 92 (juillet 1882): 2.

———. "A travers l'exposition." *La Science culinaire* 6, no. 112 (May 15, 1883a): 1–2.

———. "Exposition alimentaire: Recompenses." *La Science culinaire* 6, no. 114 (au 30 juin 1883b): 2–3.

Voltaire. "Goût." In *Dictionnaire philosophique*. Garnier, 1878. https://fr.wikisource.org/wiki/Dictionnaire_philosophique/Garnier_(1878)/Go%C3%BBt.

Zlotnick, Susan. "Domesticating Imperialism: Curry and Cookbooks in Victorian England," *Frontiers: A Journal of Women Studies* 16, no. 2/3 (1996): 51–68.

5

TASTE, OTHERWISE

From a global perspective that does not presuppose the superiority of the West, gastronomy stands out because the processes of desensualization, bureaucratization and racialization produced a peculiarly non-affective approach to taste. As I argue throughout this book, the modern notion of taste articulated in gastronomic writing is the result of how gastronomy engaged with problems specific to Western thought and with the needs of globalizing imperial colonialism and capitalism. In contrast, cultures of taste that emerged from other systems of thought and developed in different political and economic contexts have more fully embraced the subjectivity and affective power of taste. This chapter looks at how the concept of taste was conceptualized in a selection of some of the most influential systems of thought from around the world. What the selection illustrates is that each one of these systems gave serious consideration to the sense of taste in their theories of knowledge and beauty. The point is not that these systems of thought totally lacked perspectives that were distrustful of or hostile toward the senses. The point is that they do have longstanding and robust intellectual debates that value taste positively. The place of taste in the realm of knowledge and beauty has not been as consistently rejected in other systems of thought as much as it has been in the predominant strands of the Western philosophical tradition.

A decolonial perspective of the variety of ways in which humans have constructed the sense of taste across time and space needs to reject the modern Eurocentric mythology in which the particular form that any given human activity has taken in the West is the highest form that others must adopt to be considered fully developed. Philosophy, aesthetics, epistemology and gastronomy are only the contingent forms that the understanding of thought, beauty, art, knowledge and taste has taken in the West.[1] Humans everywhere had been engaged in the conceptualization of these concerns before they got their Greek names. I continue to use

DOI: 10.4324/9781003331834-6

the Greek-origin words that are the norm in the English language for the sake of clarity, but this should not be understood as accepting the notion that these practices originate in the West. My exploration of conceptualizations of taste looks into different kinds of texts, because conceptualization of taste is present in all kinds of lettered activities and is not limited to a single specific genre like gastronomy.

The following discussion is not an academic exercise of "inclusion," which would leave the narrative of gastronomic progress unchallenged. The decolonial intent of this exploration of the ways in which other peoples have conceptualized taste is to expose the fallacy of the narrative of gastronomic progress, making obvious the shortcomings of the modern approach to the experience of taste and sparking awareness of the many possibilities that the experience of taste can offer when not beholden to modern/colonial strictures. The main philosophical issue that accounts for the limited understanding of the sense of taste in Western thought is the sharp separation and hierarchy between the mind and the body and the resulting devaluation of the body and sensory experience. There have been positive valuations of sensuality in Western thought, but they have not been as influential as the idea of the body being different and inferior to the mind. Even gastronomic writers constructed their notion of taste without challenging this dichotomy. Many contemporary thinkers in the West recognize the limitations of the modern notion of taste and are working to overcome them.[2] Learning from, rather than about, the notions of taste of peoples whose systems of thought were not anchored in the separation between mind and body and therefore did not shun sensing could be a first step in the direction toward a more fully embodied and affective experience of taste in modern societies.

Theories of taste, knowledge and subjectivity that are gaining traction in Western thought are consistent with notions that were long ago developed by cultures of taste that gastronomy had supposedly superseded. Cultures of taste that did not respond to Christian thought, modern philosophy, capitalism, colonialism and modern racial thought are infinitely varied. However, most of them contain extensive elaborations of positive valuations of the sense of taste and an openness to its subjectivity and affective power. Different from the experience of the West, cultures that did not begin with a negative attitude toward the body integrated taste early on into sophisticated theories of art, beauty and knowledge. They had no need to spend centuries debating whether an art or science of taste could possibly exist. They did not need to elevate taste because they did not consider it low, to begin with. Many aspects of gastronomic culture that have been hailed as unprecedented achievements of humanity are only achievements in the provincial European context. Gastronomy does not represent the first time that humans intellectualized and aestheticized food and taste. It is also not the first time that people discussed and wrote about taste, developed and codified recipes, followed dining etiquette or indulged in culinary creativity, eating for pleasure or dining in the public sphere.

This chapter presents a discussion of a few examples of concepts of taste developed in Classical Arabic, Chinese, Indic, Nahua and Yorùbá systems of thought. Each one of these systems of thought has directly or indirectly informed many different culinary cultures across large regions for long periods of time. These cultures of taste continue to exist even when embattled by gastronomic perspectives that downgrade them as primitive, traditional or non-modern even while mining their knowledge and resources. The decolonization of taste would require not only stopping this aggression but also rejecting the modern non-affective notion of taste that has been a badge of identity of modern peoples who see themselves as belonging to superior races and civilizations.

This discussion is not meant to be exhaustive, and it does not intend to value the chosen examples more highly than any other possible ones. Each section is based on a variety of materials written by experts in each field, and readers are urged to consult the reference list to learn more. The aim of this chapter is to illustrate how taste has not been, and does not have to be, limited to the modern/colonial model constructed by gastronomic writers and which continues to inform and mediate the experience of taste in modern societies. Contrary to the myth of the gastronomic liberation of taste, the comparative decolonial analysis of a diversity of cultures of taste sketched below suggests that the cultures of taste of times and places other than the modern West have allowed for a fuller enjoyment of the pleasures of taste than gastronomy has ever dared to even suggest.

The following discussion of the conceptualizations of taste found in five different classical thought traditions cannot possibly do justice to their richness and complexity. However, even this brief review of how taste has been conceptualized for centuries in different parts of the world makes clear that the position of taste in epistemic and aesthetic thought has been more troubled in the Western tradition than elsewhere. Gastronomic thinkers in Europe took on the task of improving the valuation of taste in European lettered culture. Their struggle with a philosophical and religious tradition that shunned the senses and the body, coupled with their investment in modern racial thought, led gastronomic writers to conceptualize taste as an objective reality. Downplaying the subjective and affective aspects of taste allowed them to raise the value of taste in Western thought and also allowed them to posit the modern non-affective approach to taste as a bodily discipline that distinguished their supposedly superior civilization from all others. Learning from other conceptualizations of taste would help expand the possibilities of the experience of taste beyond the limitations of gastronomic notions. This requires renouncing the claims of racial and cultural superiority, which were constitutive of gastronomic thought.

Taste in Classical Arabic Thought

The modern notion of taste constructed by gastronomy was proposed as superior to all others, but it was shaped to a great extent as a negation of the notion of taste of the Arab and Islamic worlds. Throughout the Middle Ages, Europe looked up to Arabic thought and culture as a source of philosophical and culinary sophistication. It was only in the modern era, as Europe gained a position of geopolitical centrality through colonial incursions into other continents, that the efforts of European thinkers and gastronomers focused on distinguishing themselves from the Arabic world and attempted to erase their relationship and indebtedness to a culture they increasingly wanted to characterize as outdated. Medieval Arabic and Islamic cuisines were designed to flatter the senses with complexly spiced, aromatic and well-presented dishes. The desire to be different from the Arabic world is at the origin of the modern rejection of spices and its commitment to non-affective taste. The modern taboo against spices and the sensuous cuisine that they represent was a foundational stone in the construction of Western identity as radically different from the ones collectively defined and racialized as "Oriental."

The centuries known in European history as the Middle Ages correspond to the Classical era in the Islamic world. The authors who defined Classical Arabic thought wrote in Arabic, but they were not all ethnically Arab or Muslim. They came from many different ethnic groups and regions, and many of them were Arabic-speaking Jews and Christians.[3] Classical Arabic thought is the product of a cosmopolitan Islamic world that was excelling economically, culturally and scientifically and that had a wide geohistorical span. In his ground-breaking study of aesthetics in Arabic thought, José Miguel Puerta Vílchez demonstrates that Arab culture pioneered in creating an artistic psychology and a theory of aesthetic fulfillment, and that it transformed the arts into a complex field that included sensibility, knowledge and practice.[4] Classical Arabic aesthetic thought informed a culture of taste that acknowledged and exalted the complexity of the experience of taste as pleasure, art and knowledge. In the following paragraphs, I discuss how the pleasures of taste were asserted in religious and medical discourses before addressing how philosophers conceptualized the epistemic and aesthetic dimensions of the sense of taste and finally describing the styles of cuisine that this framework of thought enabled.

Eurocentric accounts of Arab-Islamic taste culture focus almost exclusively on Islamic food restrictions as a defining and limiting feature. This essentialist view of Islamic cultures has not allowed for a fair appreciation of the sophistication of Arab-Islamic cuisines and theories of taste. Religious thought is one of the aspects that shaped the food culture of the Islamic world, but it is not the only one. Medical, aesthetic and epistemological thought had at least as much influence in defining the development of a complex taste culture that cannot be dismissed as prisoner to religious or any single kind of thought. Furthermore, the impact that religion had on Arabic taste culture was far from being a hindrance. Whereas the Quran imposed a ban on the consumption of pork and alcohol, it also encouraged the enjoyment of the pleasures of food and taste. It has been argued that the alimentary norms laid down by Islam would have seemed like a liberation to Jews and Christians, who were required to observe many more and more cumbersome dietary rules.[5] Arabic and Islamic Studies scholar Lilia Zaouali has presented a view of Ramadan, a period during which observant Muslims fast from sunrise to sunset and break the fast with shared meals, as more than a month of fasting. She describes the Ramadan as a month of piety but also as "a gastronomic event unique in the world, lasting twenty-nine or thirty days in a row."[6] Ramadan is characterized by the enjoyment of an abundance of fine foods prepared with utmost care. An ascetic attitude toward food is incompatible with Islam because the Quran exhorts the faithful to enjoy all the pleasures that God has made available and lawful. The avoidance of such pleasures has even been regarded with suspicion, as when an eleventh-century poet and philosopher was accused of heresy because of his avoidance of meat.[7] In the context of Quranic thought, eating pleasure is not just tolerated, it is actively encouraged.

Calls to moderation and abstention from food indulgence were not absent in Classical Islamic cultures, but in general an attitude of frank delight predominated. The culinary pleasures that could be enjoyed in the world were seen as only a preview of bigger pleasures to be enjoyed in the afterlife. Good food is one of the main aspects of paradise, whose culinary delights were superlative in both quantity and quality. Details of this food utopia were richly elaborated in the Quran and in further writings, which probably influenced European fantasies about the land of Cockaigne.[8] Theologian Abu Hamid Al Ghazālī (1058–1111) went as far as affirming that in paradise people would be able to eat a hundred times more than they used to do on earth, but without suffering any ill effects.[9] Indulgence in the fantasy of a limitless capacity to eat could barely be acknowledged by European gastronomers without being accused of gluttony. But in Islamic thought, it could be entertained as a way of encouraging adherence to a religious life. Independently of this goal, the fact that the enjoyment of food was not looked down upon in Classical Islamic teachings left the door open for the elaboration of a food culture that was not hesitant to flatter the senses.

In the context of mystic thought, cooking skill and creativity were highly regarded. The thirteenth-century Islamic mystic Mevlana (1207–1273), also known as Rumi, used the language of food to illustrate his philosophy. His work is a source of information on all subjects related to the food culture of his era, including agriculture, medicine, aesthetics, flavor, table setting, entertaining, teamwork and cooking.[10] As Nevin Halici has stated, the high status of cuisine in Mevlana's period is evidenced by written rules about the organization of the kitchen, a memorial tombstone for Mevlana's cook and the fact that the training of dervishes begins in the kitchen.[11] Culinary theory and practice were seen as relevant beyond the kitchen. The respect for culinary skill and creativity made Islamic cultures a fertile ground for culinary excellence. The relatively few food restrictions in Islam are of less consequence than the idea that God made food pleasures that should be enjoyed. This allowed for the exploration of taste and the development of elaborate cuisines that flatter the sensible body without the feelings of sin and guilt predominant in Christian attitudes toward food.

The dietary system based on humoral physiology that was dominant throughout the Islamic world and Christian Europe was another realm of thought that had an impact in the shaping of the taste culture of the Islamic world. But whereas dietary concerns were considered when creating and cooking dishes, these concerns did not mean that taste and pleasure were relegated to second place. Dietary concerns in cookery books are present, but they often do not seem evident in the complex preparations whose most obvious objective is to flatter the eyes and the palate.[12] Dietary guidelines were remarkably flexible as recipes were expected to be customized to suit the seasons as well as the sex, class, physical activity level, state of health and temperament of the eaters.[13] An Andalusian cookbook of the Almohad period makes clear that food preparations should not go against the desire and pleasure of the eater.[14] People could eat the foods that were deemed unsuitable for them by just adapting the recipes. The anonymous author of the Andalusian cookbook dramatizes the openness of the dietary system: "One thing and its contrary and opponent must be mentioned, since everyone has their preferences and everything has someone who seeks and desires it."[15] Dietary guidelines were as much about preserving health as about customizing preparations to suit the needs and desires of each individual. Recipes were not to be followed dogmatically, and they allowed for constant experimentation and creativity. Medieval Arabic cookbooks often say: "Do as you please!"[16]

Spices were of great importance in the humoral dietary system, as they were the elements capable of transforming unsuitable foods into medically acceptable ones, while also modifying and enhancing their taste. According to the Andalusian cookbook, spices are the base of cooking because they diversify the culinary preparations, improve them and give them flavor, and enhance their good aspects while avoiding the bad ones.[17] Skilled spice use was essential for the transformation of food into healthy and appetizing dishes. Many natural substances like spices, herbs, leaves, seeds, berries, roots, resins and rosebuds were used to enhance the aroma, appearance and taste of the foods.[18] The cuisine of the Classical Arabic world appealed to all the senses. The concern to comply with dietary guidelines encouraged the experimentation with different combinations of foods and spices. This resulted in a wide variety of dishes that delight for reasons that go beyond their adherence to humoral theories. By encouraging the customization of dishes to suit individual humors, humoral dietetics left people free to explore the subjectivity of taste without compelling them to comply with any guidelines rigidly presented as universal.

In the case of Classical Arabic culture, religious and medical thought did not stand in the way of the exploration and gratification of the sense of taste. Epistemic and aesthetic thought, on its part, provided an ample framework that allowed for the appreciation of gustatory perception as capable of producing knowledge and art. During the Classical Arabic period, many different authors developed aesthetic theories which, in spite of their significant differences, together serve as testimony to a cultural milieu that allowed for the exploration of gustatory taste. Ibn Khaldun (1332–1406) is one of many philosophers who pondered the relationship between theory and practice and between the arts and the sciences. He considered that embodied knowledge is different from, but not inferior to, theoretical knowledge. For Ibn Khaldun, art involves learning and not just inspiration or reflection, and it also needs practice.[19] He considered that skills like cooking and singing, when learned by practice and repetition, continue to leave their mark even when a civilization is in decline.[20] It is significant that, even though the cuisine of the medieval Islamic world was thoroughly recorded in writing, Ibn Khaldun considers embodied knowledge to be most enduring.

The recognition of the cognitive and aesthetic capabilities of the body and the senses were further elaborated by Ibn Bājja (1085–1138). He divided pleasures into physical and intelligible. While food is a physical pleasure, Ibn Bājja considered that sensory pleasures were not physical but rather a low level of intelligible.[21] Because sensations for him represent an abstraction from matter, they allow for an aesthetics of the senses that is disinterested and intellectualized.[22] Distinguishing tasting from eating allowed Ibn Bājja to present a positive valuation of the sense of taste. In a similar way, Ibn Rushd (1126–1198), also known as Averroes, considered physical sensibility as a pre-rational form of knowledge as the source of aesthetic understanding.[23] Both thinkers considered the cognitive power of the senses to be inferior to reason, but they granted that each sense provided its very own kind of pathway to knowledge. Because of this, the cultivation of taste could not be devalued as a purely animal endeavor. Arabic thought recognized the epistemic and aesthetic capabilities of the senses centuries before the field of aesthetics took shape and developed similar ideas in Europe.

Mystic thinkers went even further in their recognition of the cognitive and aesthetic power of the senses. Al Ghazālī developed a metaphorical concept of taste as the highest way of knowing. For Al Ghazālī, tasting defined as a finding was superior to knowing as drawing of analogies and to faith as mere acceptance.[24] Likewise, for Ibn 'Arabī (1165–1240) the senses were not an impediment to knowledge, but they were allied to internal vision and Imagination to bring about illumination. According to Ibn 'Arabī, all the senses act with equal force and he granted them an active creative function.[25] Ibn 'Arabī's conceptualization of the senses as having equal force stands out because it does not subordinate taste as a lower sense. This notion is consistent with a culinary practice that paid attention to taste in its own right instead of having it mimic the visual arts as has often been the case in modern gastronomy. The mystics' metaphorical understanding of taste as a higher capability can only be based on a notion of gustatory taste that has cognitive, aesthetic and creative capabilities in its own right. Islamic mystics influenced the Spanish mystics Teresa of Avila and John of the Cross, who in turn influenced French thinkers in the seventeenth century. This influence allowed the French to begin the transformations in thought needed for the development of gastronomy as a legitimate pursuit of the pleasures of taste, a legitimacy that had been until then negated in European thought.[26] As we have seen, this pursuit was never illegitimate in the context of Arabic thought.

When recognizing the cognitive capabilities of taste, Arabic aesthetic thought demonstrates a complex understanding of its simultaneous objectivity and subjectivity. While modern rationalism led gastronomy to focus on the objectivity of taste and play down its subjective aspects, Arabic thought allowed for the exploration of its subjectivity. Many influential thinkers developed notions of aesthetic taste with varying levels of subjectivity. For Ibn Miskawayh (932–1030), aesthetic judgments depend on temperament and are subjective, so they "are unlimited, cannot be reduced to a technique, and have no rules."[27] Variations in the humors account for infinite variations in taste. Ibn Khaldun, when elaborating his concept of taste as applied to the art of rhetoric, expressed that taste in language can be either natural or acquired. However, the lack of the first gets in the way of the second. According to Ibn Khaldun, taste for eloquence in dialectical poetry is possessed only by those who have contact with the specific dialect in which a poem is composed and who have ample practice using it among the people who speak it.[28] While for Miskawayh perfect taste is possible but rare due to deviations in the humors, for Ibn Khaldun there is no universal perfect taste, only perfect taste in a specific context.

Ibn Rushd on his part argued that the diversity of tastes does not rule out the objectivity of perception. Healthy taste can recognize a taste just as it is, but many variables affect this perception. Because of this, sensibles are not true in themselves but depend on the perceiver and in the conditions of perception.[29] While not denying the objectivity of taste, which makes things knowable, the openness to the variability and subjectivity of taste allowed for the possibility of a culinary culture that engaged not only the natural taste of foods but also the many ways in which its perception can be varied. The complex understanding of the subjectivity of taste in Arabic thought is in many ways consistent with ideas that are presently gaining traction in Western thought, like the conceptualization of taste as an experience and as the result of the interaction between the taster and the tasted. This is more conducive to culinary experimentation than the narrow ideas espoused by modern gastronomy that were focused first and foremost on a singular and objective natural taste of each food.

Artists of all kinds were expected to take into account the variability of perception. In an epistle on music by the secret society of Muslim philosophers known as the Brethren of Purity, the authors indicate that skilled musicians will vary their melodies and adapt them for each occasion to avoid the perception fatigue and boredom that also affect food, drinks, scents and things seen and heard.[30] For the Brethren of Purity, ideal proportion is the key to artistic perfection. There should be harmony in colors, shapes and sizes and also in flavors and aromas.[31] While the principle of harmony in art is based on the harmony found in nature, authors like Abū Hayyān al Tawhīdī (923–1023) saw art as the transformation of nature by human thought, which in turn depends on temperament and cannot be explained rationally. Keeping the tension between the objectivity and subjectivity of sensory perception allowed ample scope for a notion of art that was creative and transformative. While the aesthetic theories found in Classical Arabic thought are rich and varied, for the most part they enabled a notion of art that addressed the body and the senses as capable of affecting and being affected by the objects of perception. This is consistent with an abundant repertoire of cookbooks with recipes that paid careful attention to how the dishes would engage the senses. The understanding of the simultaneous objectivity and subjectivity of taste in aesthetic thought allowed for the development of a culinary art that goes beyond any "natural" or objective taste to explore the aesthetic possibilities of gustatory perception.

It should not be surprising that an intellectual culture with such a positive and sophisticated understanding of the sense of taste produced an elaborate cuisine. Islam has the most extensive medieval food literature in the world.[32] The cookbooks and other food-related manuals do not concern themselves only with the cooking of Muslims. They contain recipes for dishes of all the peoples in the Abassid Empire, which included non-Arab Muslims and non-Muslims.[33] The texts are testimony to a flourishing cosmopolitan and refined cuisine. Medieval Arabic cookbooks have a pleasure-seeking goal that suited the urban affluent, learned and leisured classes.[34] Markets like a ninth-century food market in Baghdad offered plenty of choices including fried and roasted meats, fish and sausages, hot stews, cold vegetable dishes and over 50 sweet confectionery options.[35] The cookbooks also offered ample choices of recipes from a wide variety of places, time periods, social classes and dietary customs. The extant cookbooks range from the lavish tenth-century *Kitāp al-Tabīkh* by Ibn Sayyār al-Warrāq, which compiled and simplified many sources of high cuisine for the new rich, to the simple fifteenth-century *Kitāb al-Tibākha* by Ibn al-Mabrad, which presents abbreviated recipes for modest dishes.

Classical Arabic cookery books were part of a corpus of literature with food as its subject without any moral, ethical or religious connotations.[36] Food and drink, including wine, are central features of poems, narratives, encyclopedias and many other genres. The texts praise food, make food recommendations on taste and health grounds, discuss food preparations and dining etiquette, and ponder the relationship between words and food, or eating versus talking.[37] The texts present a wide variety of perspectives on food and use an equally varied range of styles and tones. Food was not excluded from erudite or artistic thought, like it was in Europe. Al-Warrāq's *Kitāp al-Tabīkh*, which is the oldest preserved Arabic cookbook, includes hundreds of recipes taken from cookbooks written by or for caliphs, princes, physicians, professionals, political and literary figures.[38] Aside from anthologizing and making accessible Abbasid cuisine, the book discusses kitchen equipment, ingredients, humoral qualities of foods, etiquette and advice regarding the benefits of exercising before a meal and napping afterward.[39] It also includes food-related poems and amusing anecdotes. This cookbook, and classical Arabic food-related texts in general, examine and celebrate food and its pleasures in ways that did not emerge in Europe until the nineteenth century.

Al-Warraq's cookbook exemplifies how the refined enjoyment of food was intertwined with art and erudition. The recipes are written in a clear prose that is often poetic and that evokes the sensory qualities the foods should have at different stages of cooking. A few recipes are further illustrated by verses that serve as an aid to prepare and enjoy the dish. A poem placed after a recipe for pastries describes them as follows:

> *Khushkanān* skillfully contrived. Before they were folded and sealed,
> Their shells were made into delicate thin rounds, smooth and lustrous.
> Like ornaments adorning the neck, stuffed with sugar and saffron,
> And ground almonds. The cook fried them in oil of sesame hulled,
> The way adepts masterly fry. They came out like luminous moons,
> With slender waists and pointed tips, clad in beautiful gowns.
> Like crescents outshining the night. Similar to rows of *dirhams*,
> Which a scrupulous hand has minted. Of princely honey, they had their fill.

Perfumed with excellent rose water of Jūr. Arranged thus on a crystal platter.
They do, dear folks, half moons resemble. So well made they look like
Lines upon lines of beautiful writing. As if with salt and camphor topped
The lines of sugar sprinkled look. Resplendent with their spread out gown.
They are, by God, blemish free. Our cook sent them to the vizier as a gift.[40]

The poem could serve as a recipe, but it is far from being just a utilitarian set of instructions. It praises the fineness of the ingredients and the majestic looks of their preparation and presentation with unabashed sensuality. It also hints at the connection between food and sex and between artistically displayed food and beautifully written words. The full appreciation of this poem and the food it describes requires an educated palate as much as an understanding and appreciation of cooking, literature and calligraphy.

Classical Arabic texts about food reveal a sophisticated understanding of the sense of taste consistent with the one developed in aesthetic thought. Scientists, cooks and lexicographers theorized about the different tastes, often discerning as many as nine.[41] Ibn Rushd developed a complex understanding of the senses of taste and smell, recognizing their interactions in the perception of flavors. He defined taste as the perceptive faculty of the ideas of flavors, a formulation that focuses on the role of the mind in the perception of flavor.[42] To this understanding of taste corresponds a cuisine that paid special attention to the aromas and colors of the food. Beyond the aromatics used in the process of cooking, perfumes like musk, camphor and rose water were added to dishes as part of their final preparation.[43] Natural substances like coriander, spinach, saffron, turmeric, eggs and pomegranate were used to enhance the color of the food and ensure visual appeal.[44] Dishes were compared by al-Warrāq to a garden where different varieties are combined in a beautiful and organized way.[45] This conceptualization of the art of cooking is consistent with the most characteristic view of beauty in Arab-Islamic culture, which defines beauty as the integration of individual elements to form a perfect whole.[46] This definition of beauty applies to cooking as much as it does calligraphy, goldsmithing, textiles, poetry, music or architecture. The art of cooking unquestionably belonged together with the other arts.

Classical Arabic cuisine was attentive to the affectivity of taste. Culinary art consisted of artistically playing with the variability, subjectivity and intersensoriality of taste to enhance and vary the pleasure of dining. Al-Warrāq's and other cookbooks make use of spices in many specific combinations that pursue particular effects. This contrasts with the generic use of spices in Europe even during the Middle Ages, before spices were sacrificed in pursuit of the ideal of objective, non-affective approach to taste that would characterize modern subjects.

Taste in Chinese thought

Chinese food and taste culture are often recognized in gastronomic scholarship as being highly sophisticated. Such sophistication, however, did not stop gastronomic writers from othering Chinese culture by focusing almost exclusively on details that look bizarre to Europeans, like eating bird's nests. Chinese food culture stands out for its incorporation of a remarkably large number of edible substances.[47] The knowledge of a wide range of foods and the development of cooking techniques suitable for them are the result of careful attention to their diverse organoleptic and culinary qualities. However, in gastronomic narratives, Chinese omnivorousness is routinely disparaged as the simple result of food scarcity. Obviously, not everybody in China faced food scarcity and other peoples who have faced food scarcity did not develop the same thorough knowledge of foods as the Chinese. The dismissal of omnivorousness as a sign of food scarcity demonstrates the inability of gastronomic discourse to see other taste cultures as anything other than a foil to highlight its own supposed superiority.

In China, interest in food was never considered unsuitable for rulers and thinkers. Since ancient times, food has been incorporated into cosmological, medical, aesthetic and political systems of thought. Foods were categorized according to different classifications, like the cosmic duality of yang and yin. The classification of the cosmos into five phases (earth, metal, fire, wood and water) was the base for the classification of the five flavors (sour, bitter, sweet, pungent and salt) and the five smells (rancid, scorched, fragrant, rotten and putrid).[48] Foods were also classified according to the different levels of energy (ch'i) and power (pu) that they transmit, their toxic potential (tu), and their ability to clean or dispel undesirable matters from the body (ch'ing and hsiao).[49] All these categories approach food as having distinct qualities with specific effects in the body, which are nonetheless susceptible to change according to variables in the processes of preparation and in the ingesting body. The categories took into account different kinds of effects that foods had on the body, from health-related effects to sensorial ones.

By 550 CE, Greco-Indian humoral food classifications had become dominant over Chinese ones.[50] The humoral system was compatible with Chinese thought because it stressed the Confucian values of balance, order and harmony and because the humoral cooling and heating classification worked well with the yang and yin cosmology.[51] The adoption of the humoral system in China did not lead to a focus on bodily secretions as much as it did in Europe.[52] It did not lead to the dominance of Indian spices either. The humoral system took different shapes in different times and places. Its flexible nature allowed for an engagement with foods that recognized their power to affect the human body as well as the power of food preparation and of the consuming body to affect the effects of the food. The relationship between foods, bodies and taste in the humoral system and in traditional Chinese classifications was interactive and mutually affective. E.N. Anderson has argued that Chinese traditional beliefs helped to keep the food production system diverse because it led people to explore and domesticate a wide range of plants and animals for their medical properties.[53] Chinese culinary knowledge and repertoire of dishes were also expanded because of such beliefs.

Different philosophical and religious systems of thought evoked food and taste knowledge to illustrate their points. Ancient Confucian and Daoist philosophies valued a fine-tuned palate as reflective of a well-ruled self. Confucian texts contain information on food and cooking, and Confucius has been quoted as saying "Everyone eats and drinks, but only a few appreciate flavor."[54] The celebrated French nineteenth-century differentiation between those who eat and the few that could be considered gourmands had been established in China centuries earlier. According to Siufu Tang and Isaac Yue, the quote relates to the idea that rites and norms, i.e. culture, are what give food and drink their proper taste and meaning.[55] Taste was recognized as the product of the interaction of nature (objective taste) and culture (meanings). This notion is clearly explained in the *Annals of Lü Buwei*, an encyclopedic compilation of knowledge from 239 BCE:

> It is the essential nature of the ear to desire sounds; but if the mind finds no pleasure in them, the ears will not listen even to the Five Tones. It is the essential nature of the eye to desire colors; but if the mind finds no pleasure in them, the eyes will not gaze even on the Five Colors. It is the essential nature of the nose to desire perfumed fragrances; but if the mind finds no pleasure in them, the nose will not smell them. It is the essential nature of the mouth to desire rich flavors; but if the mind finds no pleasure in them, the mouth will not taste even the Five Tastes. The locus of the desire is the ears, eyes, nose, or mouth, but the locus of pleasure or displeasure is the mind. Only when the mind has first attained harmony and equilibrium does it find pleasure in such things.[56]

While distinguishing the mind from the senses, Lü Buwei recognizes sensory pleasure as having an epistemic and aesthetic dimension. Flavors were conceptualized as the result of the collaboration of body and mind.

Other classic texts like the *Tao Te Ching* (c. 400 BCE) condemned fine dining, and the Daoist approach to food was rather ascetic.[57] In medieval China, Buddhism led to large-scale vegetarianism and to the development of new vegetarian foods.[58] However, neither asceticism nor vegetarianism obliterated the respectable position that food and tasting had in epistemological thought. A Ming short story presents a Daoist old gentleman who refused to eat the flesh-looking food that his alchemist hosts had offered him. When he was told that the food that he refused was in fact rare roots capable of imparting immortality, the old man realized that his state of enlightenment was not advanced enough to recognize the true nature of the food and therefore he did not merit immortality.[59] In this tale, discerning senses regarding food are not a minor form of knowledge but an indispensable knowledge to attain enlightenment, if not immortality.

Food knowledge was respected enough to be useful in the articulation of principles of political theory. To convey that care and attention are an important principle of governance, a Daoist text claimed that governing a country is like cooking a small fish.[60] A small fish can be delectable when handled properly, but it is quickly ruined in the hands of inexperienced cooks. Culinary and political theories are seamlessly blended into each other by the cook and political advisor

Yi Yin (1648–1549 BCE). In the course of a brief conversation reported in *The Annals of Lü Buwei*, Yi Yin lays down a theory of taste and cooking and presents it as a motivation for territorial expansion.[61] Because the perfect flavors correspond to different regions, and the most delectable delicacies span a vast geography, the only way to satisfy the desire to have them all was by extending political control. Yi Yin provoked and engaged the ruler's desire to experience and have all the tastes, transforming it into the engine for a political project of integration. If sensory knowledge and pleasure could be an engine for politics, culinary knowledge was not just a metaphor but a valid form of knowledge capable of sustaining a philosophy of governance. Once the equivalence between different flavors and different regions is established, Yi Yin's recommendation of the specific ways in which the different foods must be cooked to bring out their best flavor can be understood as suggesting that each region must be handled according to the specific traits of its people. Yi Yin's culinary and political principles illuminate each other. While the use of culinary language makes the political theory understandable, the fact that culinary thought could function in this way reveals much about a more generalized Chinese approach to taste and cooking. Yi Yin's ideas would not have resonated if the pleasures of taste were not already publicly susceptible to be accepted as a valid political motive. Similarly, the fact that governance and cooking could stand in for each other underscores a conceptualization of both cooking and tasting as active endeavors that require knowledge and skill.

Chinese literary theory also shows a complex understanding of the sense of taste. Discussions about poetry often use the word *wei*, which is usually translated as taste but carries a wider range of meanings than the English word does.[62] Tastes are few and can be listed and classified, but flavors are more elusive. Their elusiveness is exactly what turned them into a principle for poetry. Beyond structure and meaning, poetry was expected to have flavor.[63] Flavor was an indescribable and yet indispensable characteristic of good poetry that few poets could impart to their compositions and few readers or listeners could appreciate. The ineffability of taste was in part responsible for its marginalization in traditional Western thought until the formalization of the field of aesthetics. But Chinese literary theory embraced the ineffability of taste. *Wei* works as a literary principle because it was already understood in culinary thought. The Chinese concept of gustatory taste acknowledged that the flavor of food was the result of both objective and subjective variables. The Chinese concept of aesthetic taste was built on a notion of gustatory taste that acknowledged its complexity.

A classic text for the study of Chinese taste culture is the memoir of the late Ming dynasty historian Zhang Dai (1597–1689). In the transition from Ming to Qing, Zhang Dai went from having an affluent and extravagant lifestyle to not having enough to eat.[64] His memoir is a window into the refined lifestyle that the Ming elites used to have, including their enthusiasm for well-appointed food. His grandfather had formed an Eating and Drinking Society to investigate the proper tastes of things and compiled a *History of Cooked Food*.[65] Zhang Dai, on his part, had formed a Crab Society for the exclusive purpose of eating crabs seconds after being captured at the height of the season, when they most perfectly embodied the Five Flavors without the addition of any seasonings.[66] Evoking the flavors of the crabs that Dai enjoyed with his brothers establishes a sense of belonging based on a common sense of taste.[67] As the Ming elites were upstaged by new rulers and a new rich merchant class, Zhang Dai's text sought to explain the good taste that defined his class and that eluded the new rich.

The historical moment in which Zhang Dai wrote in China is similar to the moment in which Alexandre-Balthazar-Laurent Grimod de la Reynière (1758–1837) wrote in France. As members of a privileged class that had lost its position, both authors set out to write down the standards of taste that defined their class and that were all that they had left to try to secure a position in a new social, political and economic order. Both China and France have texts that explain standards of taste, but this is not necessarily proof of their having a more self-conscious taste culture than others. It is only because of the dramatic demise of a powerful class that we have written standards of taste for late Ming China and post-revolutionary France. Elsewhere, powerful classes that have been phased out less abruptly felt no need to write down their standards of taste, because they were well-known and faced less harsh challenges. While appreciating the existence of texts that spell out the standards of taste of a particular class in a specific time and place, the absence of such texts should not be taken to mean that clear and sophisticated standards of taste were absent in other periods and places.

In eighteenth-century China, there was a proliferation of texts centered around food and cooking.[68] Many scholars compiled and published recipe collections. One of the most famous collections is Yuan Mei's (1716–1798) *Recipes from the Garden of Contentment*. Yuan Mei writes from the perspective of a lettered patron of good cooking, who has access to the best foods and knows enough theory of cooking to articulate exacting standards for their preparation. It was figures similar to Yuan Mei who wrote the foundational texts of French gastronomy in the nineteenth century. Yuan Mei is an important Chinese cultural figure in both the literary and the culinary fields. He is considered one of the greatest poets of eighteenth-century China, and his work on Chinese food is revered by Chinese chefs.[69] His main contribution is not originality, but the eloquent capture in written form of the cooking procedures and standards of taste of a Chinese culinary culture that was centuries in the making.

The taste culture documented by Yuan Mei is embedded in the general aesthetic values and debates in Chinese thought. Ancient Chinese aesthetic thought focused on the sensuous nature of beauty. It affirmed the senses and the pleasures that they provide, although in a regulated fashion. As Zehou Li put it: "The Chinese aesthetic tradition emphasizes the moderation of violent sensuality, the rationality inherent in the perceptual senses, and the social inherent in the natural."[70] Chinese aesthetic thought did not draw a sharp separation or hierarchy between sensuality and reason. This allowed for the engagement with taste in all its dimensions: from the objective to the cultural and psychological. In Confucian thought, the senses and desire could be satisfied but always within a framework of moderation mediated by culture and society. The regulated validation of the senses and desire was one of the main goals of nineteenth-century French gastronomers. In China, such a validation has been a constant given. But Yuan Mei's work as a poet and theorist was written during a period in which there was a push for an even stronger assertion of the body and sensory pleasure. Aesthetic thinkers were repudiating to different degrees the controls that aimed to restrain pleasure with traditional standards, norms and aesthetic criteria.[71] Yuan Mei's compilation of recipes and standards of taste demonstrates adherence to traditional culinary and taste standards. However, he felt the need to explain and justify the culinary canons, which were being questioned by those who espoused more individualistic aesthetic approaches. In the preface, Mei explains:

> If someone says, "Everybody has their own preferences, just as they all have different faces, how can you be so sure that their tastes will match your own in any way?" To that I say, "Like arranging a marriage and chopping wood for an axe handle, if things are done in an orderly and practical manner, then the results will not be too far off the mark. I cannot guarantee that all the people under heaven will have the same tastes as I do, but I can still introduce them to dishes and recipes that I fancy.[72]

Yuan Mei grants that there are differences in taste and that no food is capable of pleasing all, but he still considers that a set of principles and procedures is useful. Principles, without being universal, can bring us closer to an agreement or at least to the understanding of what pleases others even if it does not please us. For Yuan Mei, the establishment of principles was not done with the intention of establishing the universality or superiority of his preferences, even when he expressed such preferences enthusiastically.

Chinese taste culture is based on a multidimensional notion of taste that was never seriously hindered by hostility from medical, religious, epistemological or aesthetic thought. At a superficial level, Chinese cuisine seems to share with French cuisine the ideal of respecting the "natural" taste of the foods. However, while for the French the taste of the food is understood as a simple objective fact, for the Chinese it is something considerably more complex. Each food contains different tastes, and the cook must select the appropriate ones and prepare them in a way that would be suitable for a specific culinary and social setting. Unlike nineteenth-century French gastronomy, Chinese culinary thought did not simplify the complexity of taste.

Taste in Indic Thought

Even though the taste cultures of the Indian subcontinent are notoriously complex and refined, they are routinely excluded or quickly dismissed in accounts of the global history of taste. This is the result of looking at Indian taste cultures through the distorted lens of the narrative of modern gastronomic progress. Culinary histories of India focus most of their attention on the role of Moghul and British rule in the construction of what today is known as Indian cuisine. A superficial understanding of the nature of cultural flows and interactions allows these histories to portray India as a passive receiver of cultural influences. A better understanding of cultural relocations would take into account the active role of Indians in the transformation of myriad cultural influences into new forms of culture. The dishes brought by the Moghuls from Persia, Turkey and Central Asia became noticeably more elaborate and refined in India, where they were enriched by Indian ingredients and culinary knowledge. The kebabs brought by the Moghuls, for example, are only basic skewered meats compared to the hundreds of lavish varieties of kebab preparations developed in India. What the world knows as Indian cuisine is only a fraction of the rich diversity of Indian and South Asian regional cuisines, which are the result of centuries of development and interactions. There is a growing bibliography on the cultural politics of food in India in the colonial and postcolonial contexts, which is illuminating. However, there has been very little change on how Indian thought continues to be excluded or misrepresented in global accounts of the history of taste.

The dismissal of India in global histories of taste has for too long been justified by a partial misreading of Arjun Appadurai's classic essays on Indian cuisine. While granting that the art of cooking is highly developed in India, Appadurai focused his attention on elucidating why India had a relative scarcity of cookbooks before the postcolonial era and why a national cuisine did not start to emerge until the preindustrial and postcolonial era.[73] In spite of Appadurai's nuanced assessment of how the specific conditions in which India's cuisines developed were not conducive to the writing of cookbooks and the standardization of a national cuisine, many readers took the lack of cookbooks and of a national cuisine as signs of an underdeveloped cuisine and taste culture. The written word and the nation form were essential for the development of French cuisine and modern gastronomy so, from a Eurocentric perspective, their absence becomes a defect. However, neither the written word nor the nation form should be taken as a universal or superior characteristic. It is clear that written cookbooks have not been necessary for the development, transmission, refinement and preservation of India's many sophisticated cuisines. The absence of a national cuisine is only a testimony to the cultural richness of India, which includes dozens of distinct regional cuisines.

It is considered obvious that religious constraints and Ayurvedic dietary guidelines have stood in the way of the cultivation of taste for its own sake in India. However unwittingly, Appadurai's assertion that "[f]ood thus stays encompassed within the moral and medical modes of Hindu thought, and never becomes the basis of an autonomous epicurean or gustatory logic"[74] has been considered sufficient evidence to prove this point. Appadurai's statement accepts as valid the claim that modern societies have been able to free their taste cultures from non-gustatory concerns. However, the existence of any purely gustatory approach to taste is a fallacy. Taste is always the result of complex interactions of chemistry, biology, psychology and culture. As I argue in this book, the approach to taste in modern gastronomy was determined by modern thought and its capitalist, colonial and racist ideologies. The Indian notion of taste, on its part, is embedded in classical Indian aesthetics, medicine and cosmology. In the case of modern gastronomy, the interaction of taste with different aspects of modern thought led to the abdication of the affectivity of taste. In contrast, in India the interaction of taste with other aspects of Indian thought led to a fully affective notion of taste.

Indian aesthetics, medicine and religion have served to potentiate the pleasures that can be derived from the experience of taste. Whereas in European thought taste was a latecomer, and in Arabic and Chinese thought it has always been present with varying degrees of importance, in classical Indian thought the conceptualization of taste is foundational. The remarkable scarcity of texts about culinary art in an otherwise highly textualized civilization should not be taken as proof of the subordinate status of cuisine. Quite the opposite: the development of textual discourse on aesthetics, medicine and religion depended on a stable understanding of taste and the art of cooking. Theorization about taste and cooking had to be already developed and understood by all parties before it could be used as the founding metaphor to explain other aspects of classical Indic thought. In the following paragraphs, I discuss how taste has been conceptualized in classical Indian aesthetics, Ayurvedic medicine, cookbooks, encyclopedic memoirs and Hindu thought.

Classical Indian aesthetics, as seen in Sanskrit texts from the third to the eighteenth century, was dedicated to the definition of *rasa* in literary texts, including drama and poetry. *Rasa* is usually translated as taste, but it is a more complex concept than what the English word conveys. According to Sheldon Pollock, *rasa* theory emerged and developed for fifteen centuries independently of the sphere of religion.[75] Classical Indian aesthetics used gustatory taste as a metaphor to explain the production and reception of the emotion and pleasure produced by literature. Western thought also used gustatory taste as a metaphor, but it was mostly concerned with aesthetic judgment. In classical Indian thought, the notion of gustatory taste that serves as a metaphor for aesthetic taste is not just an inferior "prosthesis to reason"[76] as in Western aesthetics but a complex affective experience in which mind and body are not separate or hierarchized. The aim of Indian aesthetics moved from the formal process through which emotion is engendered and made accessible through the literary work toward the reader's own experience of this emotion.[77] The role of the taster is important in the Indian understanding of gustatory and aesthetic taste. Unlike modern Western thought, which took gustatory taste only as a starting metaphor for a "superior" aesthetic taste, Indian classical aesthetics were modeled after the experience of taste without shunning its hard-to-systematize complexity.

One of the foundational texts of Indian classical aesthetics is the *Treatise on Drama* of Bharatamuni, c. 300 CE. This text invokes knowledge of cooking and gustatory taste as a way of explaining its theory of *rasa*. *Rasa* is not just taste but something that is relished with aesthetic pleasure. Bharatamuni explains that the production of aesthetic *rasa* comes from the combination of *bhāvas*, or emotions, in the same way that good taste is produced "through the juice produced when different spices, herbs and other articles are pressed together."[78] The notion of gustatory taste on which Bharatamuni's theory of aesthetic *rasa* is based is one in which taste is not just an objective property of foods but something that emerges in the act of combining and tasting different elements as one. Bharatamuni also explains that learned theater audiences become delighted in the same way that "people conversant with food-stuffs and consuming articles of food consisting of various things and many spices enjoy their taste."[79] This quote suggests that there was a category of discerning people "conversant with foodstuffs" who had the ability and know-how to derive pleasure from the well-combined ingredients of a dish. The gustatory pleasure that serves as a metaphor for aesthetic *rasa* was not just a bodily pleasure but one mediated by knowledge. As Sheldon Pollock explains, throughout *rasa* theory aesthetic *rasa* "can be regarded as a property of a text-object, a capacity of the reader-subject, and also a transaction between the two."[80] *Rasa* references a comprehensive notion of gustatory taste, understood as existing at once in the food, the taster and the act of tasting. The delights of *rasa* and of gustatory taste are an experience that involves the body, reason and emotion or affect in a unified way.

A comprehensive understanding of taste is also at the base of Ayurvedic medicine, whose origins are traced as far back as the second century BCE. The *Charaka Samhita* (c. 400–200 BCE), which is the extant text that has served as the enduring foundation for the practice of Ayurveda, begins with an extensive section on principles that includes a formal theorization of taste.[81] It explains that there are six basic tastes (sweet, sour, salty, pungent, bitter and astringent), which can be combined in 63 different variations that can in turn be combined to produce innumerable tastes. Taste in the *Charaka Samhita* is affected by many variables, including the taster, the preparation, the place or climate, and time. Variations in time include seasonality, time of day and stages in the life cycle of the food substances. Taste effects are experienced by the eater in different stages, from ingestion to digestion and excretion. Tastes were further classified by the qualities of roughness, hotness, unctuousness, coldness, heaviness and lightness. The minute attention to the many variables that affect taste reveals the importance of taste as an experience. Although intended as part of a manual on medicine, the first part of the *Charaka Samhita* can be seen as a guidebook for anybody interested in exploring the pleasures of taste. It could help cooks systematically experiment with combinations of the different types and qualities of taste, and it could also guide knowledgeable eaters to fully enjoy the complexity of the experience of taste.

The *Charaka Samhita* classifies foods as suitable or unsuitable to different individuals according to their physical and mental classification, but the variables are many and the system is elastic enough to take into account the eaters' preferences. The guidelines offered by the text never overrule gustatory pleasure. The text says little about the use of the culinary spices that are the trademark of Indian cuisines. This suggests that their use was more guided by gustatory than by strictly medical concerns. The kind of formally theorized knowledge about taste, cooking and manners codified in the *Charaka Samhita* is the kind of knowledge that nineteenth-century French gastronomers aspired to produce. The formal understanding of taste codified in Ayurveda informs many different regional cuisines in India, each one with its own logic and understanding of gustatory pleasure. In the case of Ayurveda, medical concerns did not stand in the way of the development of a pleasure-driven approach to taste. Instead, it served as a structural framework to guide its exploration. The *Charaka Samhita*, insofar as it codified taste principles, is a master text that connects and underlies many of the different cuisines of India and South Asia.

Many of the oldest available cookbooks in India refer to a theory of taste consistent with the one recorded in the *Charaka Samhita*. The *Pākadarpana* is an ancient cookbook attributed to mythological King Nala, whose story is narrated in the Mahabharata. The *Pākadarpana* explains that King Nala received from the son of the sun the special power to cook without fire and that he gifted humanity his cookbook with recipes and principles of cooking.[82] Regardless of who might actually have written the cookbook and when, the existence and popularity of the story itself underlie the importance that cooking has in Sanskrit culture. The cookbook classifies food preparations into semihard, soft, lickable or relishable, suckable and drinkable.[83] This classification is based on the physical qualities of the foods as experienced by the mouth when consumed. Whereas most cookbooks classify their preparations based on their main ingredient, the experience of the eater is at the center of the categories in the *Pākadarpana*. The book contains principles of taste, recipes and clear indications of what constitutes a good cook. All possible demerits of food are attributed to careless cooking, so the book warns cooks to avoid a list of mistakes. A good cook is expected to be well versed in measurement and accuracy,[84] and should have the know-how to vary the preparations by devising different combinations of substances.

The *Pākadarpana* is not meant to be exhaustive, but the cooks were expected to profit from its theoretical framework to prepare countless dishes. The art of cooking, recognized as one of the 64 arts in ancient India, was based on taste principles presented as universal, but which allowed for the infinite combination of flavors. The *Pākadarpana* defines the best cook as "one, who, having gone through the cookery very attentively and precisely from all aspects; possesses the knowledge of all sorts of cooking by heart."[85] Excellent cooks are those who have practiced enough to know the principles and preparations by heart. The development and transmission of culinary knowledge were not dependent on cookbooks but on cooking practice based on principles. The ideas and principles present in the *Pākadarpana* continue to be influential through translations into different Indian languages. The scarcity of cookbooks is not a flaw. It is the result of an understanding of taste that valued experimentation based on general principles.

The *Mānasollāsa* is an encyclopedic twelfth-century text by king Someshvara III. It contains information on all fields of knowledge, including ample details about the arts. A section called *Annabhoga* contains recipes for specific dishes, many of which can be recognized today in both name and contents, like *idli* and *dosa*.[86] The codification of dishes is apparent without the presence of cookbooks. The sixteenth-century text *Soopashastra* begins with Ayurvedic taste principles but also provides recipes for many preparations that reveal cooks and eaters were sensitive to delicate variations in the preparation of foods. For example, milk boiled down one-fourth, one-sixth and one-eighth are recognized as different preparations suitable for different dishes.[87] Similarly, the book describes multiple preparations of cream, ghee, yogurt and buttermilk. Special attention is given to how to prepare rice perfectly. The book explains which mistakes to avoid and which procedures to follow to "get rice where the grains are not broken and it will bloom and shine like a flower."[88] The exacting attention to detail in the preparation of rice to achieve the desired organoleptic qualities reveals a heightened aesthetic approach to the pleasures of food and taste. This book, unlike the earlier ones, makes use of a wide variety of condiments and spices, including many of the ones associated with contemporary Indian cooking. Spice use is highly variable across space and time in India, since each period and region has a different set of spices and a different logic guiding their use. Yet, spices have always been one of the elements that cooks are expected to combine in their dishes to produce new flavors and organoleptic effects. In the Indian context, the modern gastronomic ideal of a single spice mix for all dishes would be an appalling simplification.

In India, medicinal knowledge led to a complex understanding of taste that enabled gustatory and culinary exploration and development. The complexity of Indian cuisines is in part the result of the attention paid to the many different aspects of the experience taste. The relationship of Indian taste culture with religious thought has also been one that encourages excellent cooking and gustatory pleasure. Hindu religion is very heterogenous, and it includes widely divergent practices. In the most ascetic end of the spectrum, Hindu holy people should not cultivate their palate.[89] Even in this case, food is a central metaphor and an understanding of theories of cooking and taste is essential for the communication of religious ideals. The path to becoming righteous is explained by allegories of proper cooking procedures, while the enjoyment of religious bliss is compared to the joy of eating the most delicious foods.[90] However, the spectrum of Hindu religion encompasses many other sectarian communities in which actual gustatory enjoyment is at the front and center. In pilgrimage centers like Govardhan, Brahmin priests cook and enjoy delicious feasts.[91] There is the ideal of the *mastrāma*, a person whose carefree and sensuous lifestyle includes the enjoyment of abundant and tasty food, intoxicating drinks, song, prayer and wrestling.[92] In the case of the *mastrāma*, religious devotion is enacted by engaging and even overloading the senses rather than by suppressing them. Vaishnavism, one of the major Hindu denominations, developed its theory of emotional religious experience based on an adaptation of the rasa theory of Sanskrit poetics.[93] In this adaptation, Krishna becomes the source of *rasa*. Krishna is the object that is relished and the subject who relishes, as well as the embodiment of all moods and the giver of the experience of moods to others.[94] In the practice of offering food to Krishna, there is an economy in which sensory pleasures circulate and are transformed into the experience of pure love and ecstasy. The practice of temple food offerings is very old and widespread. Even though different Hindu denominations approach temple food offerings in their own way, in general temple cooking and the enjoyment of consecrated food as *prasād* puts food at the center of religious practice.

The practice of cooking and eating at temples gave Indian gustatory taste culture a number of characteristics that gastronomy advocated in France in the nineteenth century. These characteristics include the striving for excellence, the codification of favored recipes and the dissemination of excellent cooking beyond the circle of the aristocracy and the clergy. Temple cooking has encouraged fine cooking for centuries, given that people want to offer to the gods the finest foods possible. Many temples are famous for specific food items, which are often considered the finest example of their kind. The eating of the religious food offering, called *prasād*, exemplifies an affective gustatory experience in which the pleasure of eating is enhanced by the exalted nature of the food. Temple cooking has also played a part in the codification of recipes without cookbooks. Andrea Gutiérrez has argued that the inscriptions etched in stone on the façade of medieval temples could be seen as recipes. The inscriptions detail the quantities of the different ingredients that affluent temple donors specified should be used in a given feast or god feeding.[95] Even though these inscriptions are not recipes in the way that we conceive of them now, the absence of cooking procedures or of minor ingredients suggests that they were unnecessary in a context in which many knew how to best use the donated ingredients. Even today, the officials of the Tirumala Temple of Sri Venkateshwara affirm that the measurements that they follow for making food offerings strictly follow the ones specified in scriptures that are thousands of years old.[96] The carefully curated food of the temples is distributed among temple visitors and pilgrims who often have traveled far distances. In this way at least samples of temple food are experienced by people from different social strata and geographical locations, making the ideals of excellent cooking known beyond the limited circles of the affluent. Even if the role of temples in the formulation and dissemination of culinary ideals is limited, it is still remarkable that Hindu religious practice has been far from antagonistic to the pleasures of taste. The fact that aesthetics, medicine and religion were all based on a highly affective notion of the experience of taste is testimony to the importance that a highly developed sense of taste has for Indian culture in general. Whereas in the modern West taste has been forced to conform to the unaffective standards of the arts and sciences, in India the arts and religion are based on an unabashedly affective notion of taste.

Taste in Nahua Thought

While Eurocentric accounts of the history of taste declared the taste cultures of the Arabic, Chinese and Indic worlds as grand but outdated, they dismissed the cultures of the peoples of the Americas and Africa as primitive. The taste culture of the Nahua, and all other indigenous civilizations of the Americas, is usually acknowledged at the level of nature and not of culture. Food historians might mention the richness of the natural environment that gave the world staples like turkeys, tomatoes, potatoes, beans, vanilla, chocolate, tobacco, pumpkins, corn, chilies and avocadoes. Beyond that, commentators tend to focus on a perceived lack of animal protein and do not credit the indigenous peoples of the Americas with having transformed their foods in ways that enhanced their nutritive, culinary and aesthetic value. Centuries of colonialism largely destroyed the Nahua ways of producing, preparing and enjoying food. Colonial and modern historiography further contributed to this destruction by negating the sophistication of Nahua culture in general, in favor of a focus on the Aztec portrayed as cannibalistic savages. However, much of the taste culture survived in practice and in colonial chronicles. It was also transmitted to colonizers who depended on indigenous knowledge and labor for a long time.

There is ample written evidence of the complexity of Aztec food and taste culture in the chronicles written by colonizers in the sixteenth century. Hernán Cortés (1485–1547) praised the vast, orderly and well-stocked markets that he visited in Tenochtitlán and other major Aztec cities. He was astonished by the great variety of fruits, vegetables, fish, fowl, herbs and prepared dishes that were on offer. Cortés considered that the beauty, extension and abundance of the markets surpassed cities like Granada and Salamanca.[97] Bernal Díaz del Castillo (c. 1492–1584) also marveled at the size of the markets and how well stocked they were. He detailed the refined dinner of the monarch Moctezuma, which included above 300 kinds of dishes and ended with a cacao drink and a smoke of tobacco.[98] Aztec markets were stocked by a complex system of distribution that encompassed many regions across all of Mesoamerica.[99] A wide variety of foods from local and distant regions were used in established composed dishes that had specific names.[100] Dishes were always accompanied by condiments and sauces, and they used dozens of chile varieties.[101] Far from being just providers of the sensation of heat, chiles were and are skillfully used according to their specific flavors, aromas and textures. Aztec dishes were prepared with many different aromatic herbs, many of which we do not know today.[102] Many of the salsas and moles that are known today have been present before the Spanish colonization.[103] Each mole was a specific combination of different varieties of chiles, tomatoes, seeds and aromatic flavorings.[104] The ethnographic account written by Friar Bernardino de Sahagún (c. 1499–1590) gives the names of many dishes and lists their main ingredients and characteristics.[105] Sahagún also sheds light on the Aztec criteria for evaluating cooks and the foods sold by the specialized sellers of staples like cacao, corn and beans and prepared foods like tortillas and tamales.[106] There should be no doubt that the cuisine and dining habits of the Aztec were among the most sophisticated in the world at the time that the Spanish arrived.

The Aztec fondness for well-appointed food was disdainfully noted by Friar Diego Durán (1537–1588): "they are worse than epicureans and more sensuous."[107] Sensuality and interest in food were widely condemned by Christian European thought, but it was certainly legitimate in Aztec culture. They recognized themselves as perhaps a little too interested in food, as evidenced by an Aztec story about an expedition to climb mount Colhuacán to see goddess Coatlicue. The expeditioners couldn't climb all the way and died because they had become too heavy with cacao and other foods.[108] This story is a cautionary tale in favor of moderation, but the Aztec moderation ideal was compatible with a positive approach toward food and its pleasures. Many deities were related to food, and each one had its own festival with its specific foods. Religious practice stimulated the preparation and sharing of special dishes across socioeconomic lines. Interest in food was not considered incompatible with intellectual or aesthetic concerns. Food (*tlacualli*) was synonymous with the good and the beautiful (*cualli*).[109] Aztec cuisine embraced the sensuality and affectiveness of taste. It created dishes of complex flavors by combining and transforming ingredients through grinding, toasting and steaming to achieve variations in color, aroma, heat, texture and flavor. Contrary to the tenets of modern gastronomy, which essentialize taste and expect cooking to preserve the natural taste of food, which is conceived as singular and objective, Aztec culinary techniques are meant to transform the foods to create new taste sensations. The Aztec approach to taste was non-essentialist. This is consistent with major concepts in Aztec thought.

According to James Maffie, in Aztec metaphysics reality is characterized by becoming, not by being.[110] To exist means to change. All existing things, including humans and food, are constituted by the constant struggle and unity of antagonistic partners, which means that everything is unstable and ambiguous.[111] *Malinalli* is one of three major patterns of motion and change in which antagonistic partners unite and struggle. *Malinalli* is a twisting and spinning pattern of motion and change, which is exemplified by activities like spinning fiber into thread and cooking and digesting food. *Malinalli* activities transform things by transmitting energy between different kinds of things and between different conditions of the same thing.[112] Food preparation processes like frothing cacao and grinding are examples of *malinalli* motion-change. Aztec cuisine and taste culture were oriented by these general principles of Aztec metaphysics. Certain foodstuffs were considered antagonistic partners of other foodstuffs, like large folded tortillas and hot chile sauce.[113] Cooking conceived of as *malinalli* enabled the exploration of the different organoleptic qualities of foods to create dishes that balanced contrasting qualities. Eating was not limited to the rather naïve modern notion of just grasping an objective natural taste. Instead, taste was an experience that developed and unfolded as people ate. This experience of taste could be seen as an affective rollercoaster in which the taste of the food reveals different facets at different stages. Each mole and condiment preparation can be seen as the result of an effort to keep a balance in an experience of taste that was never static. The non-essentialist character of Aztec thought allowed for a complex and affective experience of taste and a sophisticated cuisine.

Taste in Yorùbá Thought

African culinary cultures have been the most vilified by gastronomy. Early gastronomers dismissed African cuisines as primitive and unappealing. Contemporary food writers and culinary historians tend to exclude African cuisines altogether, as if the continent had not produced anything worthy of mention. In the rare instances in which food journalists engage with African foods, they tend to place them in the context of war and famine, suggesting that Africa lacks the conditions of possibility for the creation of dishes of gastronomic merit.[114] The disrespect of African culinary cultures has been based on modern race discourses that negated the humanity and the epistemic and aesthetic capabilities of the peoples of Africa. However, any approach to the culinary cultures of Africa that is not invested in perpetuating racial hierarchies can only conclude that African peoples have developed sophisticated cuisines oriented by epistemic and aesthetic principles that allow for the pleasure-driven exploration of the experience of taste.

Tadeusz Lewicki's compilation of medieval Arabic sources that mention African food reveals that West Africans developed food processing and cooking techniques for a wide variety of foodstuffs.[115] European travelers from as early as the seventeenth century were amazed by excellent food preparations like lamb and nut meatballs served in a white sauce, and by the magnificence of the ceremonies surrounding food service.[116] They were also astonished by the sensuality of a feast offered by Queen Taytu Bitul in Addis Ababa in 1887.[117] Accounts like these give only a glimpse of African culinary and taste cultures in different times and places. Additional evidence of the sophistication of African cuisines can be appreciated in the way that enslaved Africans commanded kitchens in the Americas, ultimately creating new cuisines based on American, European and African foods and culinary knowledges. The cuisines created by Africans in the Americas should not be taken to be the same as the cuisines that existed and exist in Africa.[118] But they are certainly a testimony to the vitality of the culinary skills and aesthetic sensibility that enslaved Africans brought with them. African aesthetic principles survived and endured in the Americas, and there is no doubt that they are at the root of African American dance, music, cuisine, art and design.[119] Indeed, Africans were protagonists in the creation of "the stereophonic, bilingual or bifocal cultural forms originated by, but no longer the exclusive property of, blacks dispersed within the structures of feeling, producing, communicating and remembering" that Paul Gilroy called the black Atlantic world.[120]

As an example of how to begin to understand and appreciate African cuisines and taste cultures in a decolonial way, I will frame Yorùbá cuisine in the larger context of Yorùbá culture. As seen in the above examples of Arabic, Chinese, Indic and Nahua thought, Yorùbá thought has informed and encouraged the cultivation of taste. Yorùbá religious beliefs do not antagonize the enjoyment of sensual pleasures. The deities known as Òrìṣàs have their own preferred foods, which are cooked with great care and offered to them in ceremonies. The foods and methods of preparation favored by the Òrìṣàs are now spread all over the world, as Yorùbá religious practices spawned new religions like Santería and Candomblé. Like Yorùbá religious thought, Yorùbá epistemology and ethics did not stand in the way of the development of a pleasure-driven culture of taste. The oral literary corpus known as Ifá is a source of Yorùbá philosophy. Based on this corpus, Omotade Adegbindin concluded that an epistemological relativism in which knowledge is relative in time and place is an integral part of the Yorùbá thought system.[121] This epistemological openness is essential for the cultivation of a sense of taste that acknowledges both the objective and the subjective aspects of the experience of taste. Yorùbá ethical thought, for its part, envisioned the good life as a synthesis of hedonism, egoism and perfectionism.[122] This synthesis allowed for the enjoyment of personal comfort as long as you had good character and moral rectitude. There was no religious or ethical condemnation of the pleasures of food and taste in Yorùbá culture.

Robert Farris Thompson called a moral/aesthetic attitude prevalent in West Africa and in the Black Americas an "aesthetic of the cool" and defined it as an artistic interweaving of responsibility and pleasure.[123] According to Thompson, coolness "imparts order not through ascetic subtraction of body from mind, or brightness of cloth from serious endeavor, but, quite the contrary, by means of ecstatic unions of sensuous pleasure and moral responsibility."[124] Aesthetic acts like wearing brilliant clothes and eating sumptuous foods can help a person regain their balance if they have strayed away from their originary state of nobility.[125] The enjoyment of sensuous foods is not only not at odds with a heightened sense of moral responsibility, it could be one of the paths to it.

Yorùbá cuisine and taste culture echo Yorùbá aesthetic principles. Two of the most valued aesthetic principles in Yorùbá sculpture are shining smoothness and roundness.[126] These qualities are an indication of the skill and care of the artist. It is possible to see a connection with this artistic preference in the preparation of West African swallows, like pounded yam and fufu. Swallow is a category of West African foods that can be made with a wide variety of ingredients, including yams, plantains, maize, wheat and rice. What turns these staple ingredients into a swallow is the pounding and/or cooking to form a rounded mass that is used to scoop and swallow, rather than chew, soups and stews. The consistency of the swallow has to be smooth for it to glide pleasurably when swallowing. Swallows often look like a beautiful shiny ball that seems luminous sitting next to or inside a soup that can have deep green or red colors. A dish of swallow and soup has a striking visual appeal, and it provides tactile and gustatory pleasure as the swallow is molded with fingers to better carry the stew to the mouth. Outsiders often dismiss West African cuisines as being mostly plantains and root vegetables with soups, but this assessment is as misguided as dismissing French cuisine as just meats with sauces, or Italian cuisine as just pasta. Beyond their defining consistency, swallows have infinite variations depending on the main ingredient and method of preparation, which might include fermentation. These days there are even vegetable-based swallows for those who want to reduce their carbohydrate intake. Plantains and root vegetables are also used in other preparations, not only in swallow form. Plantain and root vegetable-based cuisines have as much variety as wheat, rice and corn-based ones. People choose to eat plantains, root vegetables and swallows out of gustatory and aesthetic preference.

Like swallows, Yorùbá soups and stews are frequently misunderstood or dismissed as being the result of scarcity rather than culinary choice. With swallows, soups are a central category of Yorùbá cuisine. The definition of soup is quite elastic, as it includes preparations of consistencies that range from very liquid to almost dry. The soups are made from many different ingredients in established combinations of meats, fish and numerous condiments that provide texture and aroma. A well-made soup has skillfully layered flavors and the appropriate consistency and texture for each specific variety. Soups and swallows are eaten together and there are established criteria for pairing them harmoniously. In spite of the complexity and variety of swallows and soups, they are put down by outsiders as smelly, starchy and gluey. Mucilaginous texture is unappealing to many people, particularly in the West, but the mouthfeel of this texture is prized by many other peoples around the world and Yorùbá cooks excel at producing it. Contemporary food writer Yemisí Aríbisálà is keenly aware of how badly the world speaks of Nigerian and other African cuisines. She explains that Nigerians themselves do not speak or write about their food as much as other peoples do. She says that "Nigerians have kept words away from describing food, as if speaking would diminish it."[127] Unfortunately, in a global culinary order dominated by written gastronomic discourse, the lack of written sources is taken to mean lack of sophistication. But, fortunately, Aríbisálà writes in a confident voice that refuses to be weighed down by racist Eurocentric narratives.

Notes

1 For a discussion of this problem, see Walter D. Mignolo, "The Geopolitics of Knowledge and the Colonial Difference," *The South Atlantic Quarterly* 101, no. 1 (Winter 2002): 57–96.
2 Examples of contemporary thinkers who are revaluing eating and tasting inside the Western philosophical tradition are Raymond D. Boisvert and Lisa M. Heldke, *Philosophers at Table: On Food and Being Human*, 2016; and Annemarie Mol, *Eating in Theory* (Durham, NC; London: Duke University Press, 2021).
3 Jon McGinnis and David C. Reisman, eds., *Classical Arabic Philosophy: An Anthology of Sources* (Indianapolis: Hackett Pub. Company, 2007), xiv.
4 José Miguel Puerta Vílchez, *Aesthetics in Arabic Thought: From Pre-Islamic Arabia through al-Andalus*, trans. Consuelo López-Morillas (Leiden; Boston: Brill, 2017), 848.
5 Lilia Zaouali, *Medieval Cuisine of the Islamic World: A Concise History with 174 Recipes*, trans. M.B. De Bevoise (Berkeley, CA: University of California Press, 2009), 30.
6 Zaouali, 36.
7 Zaouali, 30.
8 G.J.H. van Gelder, *God's Banquet: Food in Classical Arabic Literature* (New York: Columbia University Press, 2000), 22–23.
9 Gelder, 23.
10 Nevin Halici, *Sufi Cuisine* (London: Saqi, 2005), 18.
11 Halici, 18.
12 Bernard Rosenberger, "Dietética y cocina en el mundo musulmán occidental según el Kitab-al-Tabij, Recetario de época almohade," in *Cultura alimentaria Andalucía-América*, ed. Antonio Garrido Aranda (México: Universidad Nacional Autonóma de México, 1996), 14.
13 Rosenberger, 47.
14 Rosenberger, 55.
15 Quoted in Rosenberger, 21.
16 Zaouali, 59.
17 Rosenberger, 40.
18 Zaouali, 52–53.
19 Puerta Vílchez, 443.
20 Puerta Vílchez, 443.
21 Puerta Vílchez, 647–648.
22 Puerta Vílchez, 648.
23 Puerta Vílchez, 672.
24 Puerta Vílchez, 753.
25 Puerta Vílchez, 792.

26 For a discussion see Viktoria von Hoffmann, *From Gluttony to Enlightenment: The World of Taste in Early Modern Europe* (Urbana, IL: University of Illinois Press, 2016).
27 Puerta Vílchez, 211.
28 Puerta Vílchez, 477.
29 Puerta Vílchez, 662.
30 Puerta Vílchez, 177 n.231.
31 Puerta Vílchez, 181–182.
32 Zaouali, ix.
33 Zaouali, xxiii.
34 Manuela Marín, "Beyond Taste: The Complements of Colour and Smell in the Medieval Arab Culinary Tradition," in *A Taste of Thyme: Culinary Cultures of the Middle East*, ed. Sami Zubaida and Richard Tapper (London; New York: Tauris Parke Paperbacks, 2000), 205.
35 Charles Perry, "What to Order in Ninth-Century Baghdad," in *Medieval Arab Cookery*, ed. Maxime Rodinson and A.J. Arberry (Devon, England: Prospect Books, 2006), 219.
36 Gelder, 39.
37 For a detailed discussion of food in classical Arabic literature, see Gelder.
38 Nawal Nasrallah, "Introduction," in *Annals of the Caliphs' Kitchens: Ibn Sayyār al-Warrāq's Tenth-Century Baghdadi Cookbook* (Leiden; Boston: Brill, 2007), 13.
39 Nasrallah, 12–13.
40 K.K.A. Öhrnberg, *Annals of the Caliphs' Kitchens: Ibn Sayyār al-Warrāq's Tenth-Century Baghdadi Cookbook* (Leiden; Boston: Brill, 2007), 421.
41 Gelder, 123.
42 Puerta Vílchez, 665.
43 Marín, 208.
44 Zaouali, 53.
45 Quoted in Marín, 214.
46 Puerta Vílchez, 480.
47 Frederick J. Simoons, *Food in China: A Cultural and Historical Inquiry* (Boca Raton: CRC Press, 1991), 15.
48 E.N. Anderson, *The Food of China* (New Haven: Yale University Press, 1988), 231.
49 Anderson, 230–240.
50 Anderson, 235.
51 Anderson, 235.
52 Anderson, 235.
53 Anderson, 242.
54 Hsiang-ju Lin, *Slippery Noodles: A Culinary History of China* (London: Prospect Books, 2015), 53–55.
55 "Food and the Literati: The Gastronomic Discourse of Imperial Chinese Literature," in *Scribes of Gastronomy: Representations of Food and Drink in Imperial Chinese Literature*, eds. Isaac Yue and Siufu Tang (Hong Kong: Hong Kong University Press, 2013), 4.
56 John Knoblock, and Jeffrey K. Riegel, eds., *The Annals of Lü Buwei =: [Lü Shi Chun Qiu]: A Complete Translation and Study* (Stanford, CA: Stanford University Press, 2000), 142–143.
57 Anderson, 36.
58 Anderson, 66.
59 Kwang-chih Chang and E.N. Anderson, eds., *Food in Chinese Culture: Anthropological and Historical Perspectives* (New Haven, CT: Yale University Press, 1977), 242.
60 Joanna Waley-Cohen, "The Quest for Perfect Balance: Taste and Gastronomy in Imperial China," in *Food: The History of Taste*, ed. Paul Freedman (Berkeley: University of California Press, 2007), 101.
61 Knoblock and Riegel, 306–311.
62 Eugene Eoyang, "Beyond Visual and Aural Criteria: The Importance of Flavor in Chinese Literary Criticism," *Critical Inquiry* 6, no. 1 (1979): 100.
63 Eoyang, 103.
64 Waley-Cohen, 119.

65 Duncan Campbell, "The Obsessive Gourmet: Zhang Dai on Food and Drink," in *Scribes of Gastronomy: Representations of Food and Drink in Imperial Chinese Literature*, eds. Isaac Yue and Siufu Tang (Hong Kong: Hong Kong University. Press, 2013), 93.

66 Campbell, 90.

67 Philip Alexander Kafalas, *In Limpid Dream: Nostalgia and Zhang Dai's Reminiscences of the Ming*, Signature Books (Norwalk: East Bridge, 2007), 31.

68 Waley-Cohen, 128.

69 Nicole Mones, "Foreword," in *Recipes from the Garden of Contentment: Yuan Mei's Manual of Gastronomy*, trans. Sean J.S. Chen. (Great Barrington, MA: Berkshire Publishing Group, 2018), xix.

70 Zehou Li. *The Chinese Aesthetic Tradition*, trans. Maija Bell Samei (Honolulu: University of Hawai'i Press, 2010), 10.

71 Li, 202.

72 Mei Yuan, *Recipes from the Garden of Contentment: Yuan Mei's Manual of Gastronomy: Suiyuan Shidan*, trans. Sean J.S. Chen (Great Barrington, MA: Berkshire Publishing Group, 2018), 5.

73 Arjun Appadurai, "How to Make a National Cuisine: Cookbooks in Contemporary India," *Comparative Studies in Society and History* 30, no. 1 (January 1988): 3–24.

74 Appadurai, 11.

75 Sheldon I. Pollock, ed., *A Rasa Reader: Classical Indian Aesthetics* (New York: Columbia University Press, 2016), 21.

76 Terry Eagleton, *The Ideology of the Aesthetic* (Wiley-Blackwell, 1991), 16.

77 Pollock, 16–17.

78 Bharatamuni, *The Nātya Śāstra of Bharatamuni*, trans. Board of Scholars (Delhi: Sri Satguru Publications, 1987), 73.

79 Bharatamuni, 74.

80 Pollock, 26.

81 Gabriel van Loon, ed., *Charaka Samhita: Handbook on Ayurveda* (Morrisville, NC: Lulu, 2002).

82 Dr. Madhulika *Pākadarpaṇa of Nala: Text and English Translation with Critical Notes* (Varanasi: Chaukhambha Orientalia 2013), 7.

83 Madhulika, 2.

84 Madhulika, 23.

85 Madhulika, 78.

86 G.K. Shrigondekar, ed., *Mānasollāsa of King Somesvara*, vol. II (Baroda: Oriental Institute, 1939), 21.

87 N.P. Bhat, and Nerupama Y. Modwel eds. *Culinary Traditions of Medieval Karnataka: The Soopa Shastra of Mangarasa III*, trans. Madhukar Konantambigi (Delhi: Intangible Cultural Heritage Division, Indian National Trust for Art and Culture Heritage and B.R. Pub. Corp., 2012), 21.

88 Bhat and Modwel, 7.

89 R.S. Khare, "Food with Saints: An Aspect of Hindu Gastrosemantics," in *The Eternal Food: Gastronomic Ideas and Experiences of Hindus and Buddhists*, ed. R.S. Khare (Albany: State University of New York Press, 1992), 30.

90 Khare, 40–41.

91 Paul M. Toomey, "Mountain of Food, Mountain of Love: Ritual Inversion in the Annakūta Feast at Mount Govardhan," in *The Eternal Food: Gastronomic Ideas and Experiences of Hindus and Buddhists*, ed. R.S. Khare (Albany: State University of New York Press, 1992), 117.

92 Toomey, 118.

93 Paul M. Toomey, "Krishna's Consuming Passions: Food as Metaphor and Metonym for Emotion at Mount Govardhan," in *Religion and Emotion: Approaches and Interpretations*, ed. John Corrigan (Oxford; New York: Oxford University Press, 2004), 228.

94 Toomey, "Krishna's Consuming Passions," 228.

95   Andrea Gutiérrez, "Jewels Set in Stone: Hindu Temple Recipes in Medieval Cōḻa Epigraphy," *Religions* 9, no. 9 (September 10, 2018): 270, https://doi.org/10.3390/rel9090270.

96   A.V. Ramana Dikshitulu and Kota Neelima, *Tirumala: Sacred Foods of God* (New Delhi: Lustre Press/Roli Books, 2016), 72.

97   Hernán Cortés, "The Second Letter," in *Letters from Mexico*, trans. A.R. Pagden (New York: Grossman, 1971), 47–159.

98   Bernal Díaz del Castillo, *The Memoirs of the Conquistador Bernal Diaz Del Castillo Written by Himself Containing a True and Full Account of the Discovery and Conquest of Mexico and New Spain*, trans. John Ingram Lockhart, vol. I (London: Hatchard, 1844), 229–244.

99   Yoko Sugiura Yamamoto and Fernán González de la Vara, *La cocina mexicana a través de los siglos. México antiguo*, vol. I (México, D.F.: Ediciones Clío, 1996), 46–49.

100 Lucía Rojas de Perdomo, *Cocina Prehispánica* (Santafé de Bogotá: Voluntad, 1994), 78.

101 Fernán González de la Vara, *La cocina mexicana a través de los siglos. Epoca Prehispánica*, vol. II (México, D.F.: Ediciones Clío, 1996), 49.

102 Rojas de Perdomo, *Cocina Prehispánica*, 44.

103 González de la Vara, 50.

104 Of course, many more mole varieties have been created throughout the centuries. As Rachel Laudan has argued, many of them were Islamic dishes re-created in the colonial kitchens of Mexico. Rachel Laudan, "The Mexican Kitchen's Islamic Connection," *ARAMCO World: Arab and Islamic Cultures and Connections* 55, no. 3 (June 2004), https://archive.aramcoworld.com/issue/200403/the.mexican.kitchen.s.islamic.connection.htm. Accessed June, 2022.

105 Fray Bernardino de Sahagún, *Historia general de las cosas de la nueva España*, ed. Juan Carlos Temprano, vol. II (Madrid: Dastin, 2001), 658–663.

106 Sahagún, 789–799.

107 Diego Durán, *Historia de las Indias de Nueva España y islas de Tierra Firme*, vol. II (Mexico: Imprenta de Ignacio Escalante, 1880), 283, http://archive.org/details/historiadelasind02dur. Accessed June, 2022. My translation.

108 González de la Vara, 6–7.

109 González de la Vara, 6.

110 James Maffie, *Aztec Philosophy: Understanding a World in Motion* (Boulder, CO: University Press of Colorado, 2013), 12.

111 Maffie, 13.

112 Maffie, 14.

113 Maffie, 149.

114 Naa Baako Ako-Adjei, "How Not to Write About Africa," *Gastronomica* 15, no. 1 (February 1, 2015): 47, https://doi.org/10.1525/gfc.2015.15.1.44.

115 Tadeusz Lewicki, *West African Food in the Middle Ages: According to Arabic Sources* (Cambridge: Cambridge University Press, 2008).

116 Jessica B. Harris, *High on the Hog: A Culinary Journey from Africa to America* (New York: Bloomsbury, 2011), 11–14.

117 James McCann, *Stirring the Pot: A History of African Cuisine*, Africa in World History (Athens, OH: Ohio University Press, 2009), 65–77.

118 Ako-Adjei, 48.

119 John Thornton, *Africa and Africans in the Making of the Atlantic World, 1400–1800* (Cambridge: Cambridge University Press 1998), 221.

120 Paul Gilroy. *The Black Atlantic: Modernity and Double Consciousness*, 3. impr., reprint (London: Verso, 2002), 3.

121 Omotade Adegbindin, *Ifa in Yoruba Thought System*, African World Series (Durham, NC: Carolina Academic Press, 2014), 233.

122 Adegbindin, 145.

123 Robert Farris Thompson, "An Aesthetic of the Cool," *African Arts* 7, no. 1 (Autumn 1973): 41.
124 Thompson, 42.
125 Thompson, 42.
126 Robert Farris Thompson, "Yoruba Artistic Criticism," in *The Traditional Artist in African Societies*, ed. Warren L. D'Azevedo (Bloomington, IN: Indiana University Press, 1989), 18–61.
127 Yemisí Aríbisálà, *Longthroat Memoirs: Soups, Sex and Nigerian Taste Buds* (Abuja; London: Cassava Republic Press, 2016), 242.

References

Adegbindin, Omotade. *Ifa in Yoruba Thought System*. African World Series. Durham, NC: Carolina Academic Press, 2014.

Ako-Adjei, Naa Baako. "How Not to Write About Africa." *Gastronomica* 15, no. 1 (February 1, 2015): 44–55. https://doi.org/10.1525/gfc.2015.15.1.44.

Anderson, E.N. *The Food of China*. New Haven, CT: Yale University Press, 1988.

Appadurai, Arjun. "How to Make a National Cuisine: Cookbooks in Contemporary India." *Comparative Studies in Society and History* 30, no. 1 (January 1988): 3–24.

Aríbisálà, Yemisí. *Longthroat Memoirs: Soups, Sex and Nigerian Taste Buds*. Abuja; London: Cassava Republic Press, 2016.

Bharatamuni. *The Nātya Śāstra of Bharatamuni*. Translated by a Board of Scholars. Delhi: Sri Satguru Publications, 1987.

Bhat, N.P., and Nerupama Y. Modwel eds. *Culinary Traditions of Medieval Karnataka: The Soopa Shastra of Mangarasa III*. Translated Madhukar Konantambigi. Delhi: Intangible Cultural Heritage Division, Indian National Trust for Art and Culture Heritage and B.R. Pub. Corp., 2012.

Boisvert, Raymond D., and Lisa M. Heldke. *Philosophers at Table: On Food and Being Human*. London: Reaktion Books, 2016.

Campbell, Duncan. "The Obsessive Gourmet: Zhang Dai on Food and Drink." In *Scribes of Gastronomy: Representations of Food and Drink in Imperial Chinese Literature*. Edited by Isaac Yue and Siufu Tang, 87–96. Hong Kong: Hong Kong University Press, 2013.

Chang, Kwang-chih, and E.N. Anderson, eds. *Food in Chinese Culture: Anthropological and Historical Perspectives*. New Haven, CT: Yale University Press, 1977.

Cortés, Hernán. "The Second Letter." In *Letters from Mexico*. Translated by A.R. Pagden, 47–159. New York: Grossman, 1971.

Díaz del Castillo, Bernal. *The Memoirs of the Conquistador Bernal Diaz Del Castillo Written by Himself Containing a True and Full Account of the Discovery and Conquest of Mexico and New Spain*. Translated by John Ingram Lockhart. Vol. I. London: Hatchard, 1844.

Durán, Diego. *Historia de las Indias de Nueva España y islas de Tierra Firme*. Vol. II. Mexico: Imprenta de Ignacio Escalante, 1880. http://archive.org/details/historiadelasind02dur. Accessed June, 2022.

Eagleton, Terry. *The Ideology of the Aesthetic*. Oxford; Cambridge, MA: Blackwell, 1991.

Eoyang, Eugene. "Beyond Visual and Aural Criteria: The Importance of Flavor in Chinese Literary Criticism." *Critical Inquiry* 6, no. 1 (1979): 99–106.

Gelder, G.J.H. van. *God's Banquet: Food in Classical Arabic Literature*. New York: Columbia University Press, 2000.

Gilroy, Paul. *The Black Atlantic: Modernity and Double Consciousness*. 3. impr., Reprint. London: Verso, 2002.

González de la Vara, Fernán. *La cocina mexicana a través de los siglos. Epoca Prehispánica.* Vol. II. México, D.F.: Clío, 1996.

Gutiérrez, Andrea. "Jewels Set in Stone: Hindu Temple Recipes in Medieval Cōḷa Epigraphy." *Religions* 9, no. 9 (September 10, 2018): 270. https://doi.org/10.3390/rel9090270.

Halici, Nevin. *Sufi Cuisine.* London: Saqi, 2005.

Harris, Jessica B. *High on the Hog: A Culinary Journey from Africa to America.* New York: Bloomsbury, 2011.

Hoffmann, Viktoria von. *From Gluttony to Enlightenment: The World of Taste in Early Modern Europe.* Urbana, IL: University of Illinois Press, 2016.

Kafalas, Philip Alexander. *In Limpid Dream: Nostalgia and Zhang Dai's Reminiscences of the Ming.* Norwalk: Signature Books, 2007.

Khare, R.S. "Food with Saints: An Aspect of Hindu Gastrosemantics." In *The Eternal Food: Gastronomic Ideas and Experiences of Hindus and Buddhists.* Edited by R.S. Khare, 27–52. Albany: State University of New York Press, 1992.

Knoblock, John, and Jeffrey K. Riegel, eds. *The Annals of Lü Buwei =: [Lü Shi Chun Qiu]: A Complete Translation and Study.* Stanford, CA: Stanford University Press, 2000.

Laudan, Rachel. "The Mexican Kitchen's Islamic Connection." *ARAMCO World: Arab and Islamic Cultures and Connections* 55, no. 3 (June 2004). https://archive.aramcoworld.com/issue/200403/the.mexican.kitchen.s.islamic.connection.htm. Accessed June, 2022.

Lewicki, Tadeusz. *West African Food in the Middle Ages: According to Arabic Sources.* Cambridge: Cambridge University Press, 2008.

Li, Zehou, *The Chinese Aesthetic Tradition.* Translated by Maija Bell Samei. Honolulu: University of Hawai'i Press, 2010.

Lin, Hsiang-ju. *Slippery Noodles: A Culinary History of China.* London: Prospect Books, 2015.

Loon, Gabriel van, ed. *Charaka Samhita: Handbook on Ayurveda.* Morrisville, NC: Lulu, 2002.

Madhulika, Dr. *Pākadarpaṇa of Nala: Text and English Translation with Critical Notes.* Varanasi: Chaukhambha Orientalia, 2013.

Maffie, James. *Aztec Philosophy: Understanding a World in Motion.* Boulder, CO: University Press of Colorado, 2013.

Marín, Manuela. "Beyond Taste: The Complements of Colour and Smell in the Medieval Arab Culinary Tradition." In *A Taste of Thyme: Culinary Cultures of the Middle East.* Edited by Sami Zubaida and Richard Tapper, 205–214. London; New York: Tauris Parke Paperbacks, 2000.

McCann, James. *Stirring the Pot: A History of African Cuisine.* Athens, OH: Ohio University Press, 2009.

McGinnis, Jon, and David C. Reisman, eds. *Classical Arabic Philosophy: An Anthology of Sources.* Indianapolis, IN: Hackett Pub. Company, 2007.

Mignolo, Walter D. "The Geopolitics of Knowledge and the Colonial Difference." *The South Atlantic Quarterly* 101, no. 1 (Winter 2002): 57–96.

Mol, Annemarie. *Eating in Theory.* Durham, N.C.; London: Duke University Press, 2021.

Mones, Nicole. "Foreword." In *Recipes from the Garden of Contentment: Yuan Mei's Manual of Gastronomy.* Translated by Sean J.S. Chen, xix–xxiii. Great Barrington, MA: Berkshire Publishing Group, 2018.

Nasrallah, Nawal. "Introduction." In *Annals of the Caliphs' Kitchens: Ibn Sayyār al-Warrāq's Tenth-Century Baghdadi Cookbook,* 1–64. Leiden; Boston: Brill, 2007.

Öhrnberg, K.K.A. *Annals of the Caliphs' Kitchens: Ibn Sayyār al-Warrāq's Tenth-Century Baghdadi Cookbook.* Leiden; Boston: Brill, 2007.

Perry, Charles. "What to Order in Ninth-Century Baghdad." In *Medieval Arab Cookery*. Edited by Maxime Rodinson and A.J. Arberry, 219–223. Devon, England: Prospect Books, 2006.

Pollock, Sheldon I., ed. *A Rasa Reader: Classical Indian Aesthetics*. New York: Columbia University Press, 2016.

Puerta Vílchez, José Miguel. *Aesthetics in Arabic Thought: From Pre-Islamic Arabia through al-Andalus*. Translated by Consuelo López-Morillas. Leiden; Boston: Brill, 2017.

Ramana Dikshitulu, A.V., and Kota Neelima. *Tirumala: Sacred Foods of God*. New Delhi: Lustre Press/Roli Books, 2016.

Rojas de Perdomo, Lucía. *Cocina Prehispánica*. Santafé de Bogotá: Voluntad, 1994.

Rosenberger, Bernard. "Dietética y cocina en el mundo musulmán occidental según el Kitab-al-Tabij, Recetario de época almohade." In *Cultura alimentaria Andalucía-América*. Edited by Antonio Garrido Aranda, 13–55. México: Universidad Nacional Autonóma de México, 1996.

Sahagún, Fray Bernardino de. *Historia general de las cosas de la nueva España*. Edited by Juan Carlos Temprano. Vol. II. Madrid: Dastin, 2001.

Shrigondekar, G.K., ed. *Mānasollāsa of King Somesvara*. Vol. II. Baroda: Oriental Institute, 1939.

Simoons, Frederick J. *Food in China: A Cultural and Historical Inquiry*. Boca Raton: CRC Press, 1991.

Sugiura Yamamoto, Yoko, and Fernán González de la Vara. *La cocina mexicana a través de los siglos. México antiguo*. Vol. I. México: Ediciones Clío, 1996.

Thompson, Robert Farris. "An Aesthetic of the Cool." *African Arts* 7, no. 1 (Autumn 1973): 40–43+64–67+89–91.

———. "Yoruba Artistic Criticism." In *The Traditional Artist in African Societies*. Edited by Warren L. D'Azevedo, 18–61. Bloomington: Indiana University Press, 1989.

Thornton, John. *Africa and Africans in the Making of the Atlantic World, 1400–1800*. Cambridge: Cambridge University Press, 1998.

Toomey, Paul M. "Mountain of Food, Mountain of Love: Ritual Inversion in the Annakūta Feast at Mount Govardhan." In *The Eternal Food: Gastronomic Ideas and Experiences of Hindus and Buddhists*. Edited by R.S. Khare, 117–146. Albany: State University of New York Press, 1992.

———. "Krishna's Consuming Passions: Food as Metaphor and Metonym for Emotion at Mount Govardhan." In *Religion and Emotion: Approaches and Interpretations*. Edited by John Corrigan. Oxford; New York: Oxford University Press, 2004.

Waley-Cohen, Joanna. "The Quest for Perfect Balance: Taste and Gastronomy in Imperial China." In *Food: The History of Taste*. Edited by Paul Freedman, 99–132. Berkeley: University of California Press, 2007.

Yuan, Mei. *Recipes from the Garden of Contentment: Yuan Mei's Manual of Gastronomy: Suiyuan Shidan*. Translated by Sean J.S. Chen. Great Barrington, MA: Berkshire Publishing Group, 2018.

Yue, Isaac and Siufu Tang, eds. "Food and the Literati: The Gastronomic Discourse of Imperial Chineses Literature." In *Scribes of Gastronomy: Representations of Food and Drink in Imperial Chinese Literature*. Edited by Isaac Yue and Siufu Tang, 1–13. Hong Kong: Hong Kong University Press, 2013.

Zaouali, Lilia. *Medieval Cuisine of the Islamic World: A Concise History with 174 Recipes*. Translated by M.B. De Bevoise Berkeley, CA: University of California Press, 2009.

CONCLUSION

The Gustatory Logic of Consumer Capitalism

Gastronomic writing defined an approach to taste that has its origins in the particular philosophical and political context of nineteenth-century Europe. This approach was proposed and has been normalized as the destiny of modern societies. Gastronomy increasingly desensualized, bureaucratized and racialized taste. These processes produced an approach to taste befitting capitalist patterns of consumption. The gustatory logic that underlies the food cultures of sectors of contemporary societies that see themselves as modern originated in nineteenth-century gastronomic thought, but it has continued to develop and adapt to new manifestations of capitalism and coloniality. This chapter discusses how the processes of desensualizing, bureaucratizing and racializing taste have intensified in our times.

Modern taste, as constructed in gastronomic writing, underwrites a capitalist form of gluttony. Satisfying modern taste requires the continuous growth of a global food production system that is unsustainable. Depletion of oil, water and soil resources, environmental degradation, exploitative labor practices, extinction and privatization of biodiversity, unfair food distribution and health concerns that range from antibiotic resistance to diet-related chronic diseases like diabetes and heart disease are only a few of the problems of the modern/colonial global food system. One of the challenges in tackling these problems is the unwillingness of consumers to give up their ability to have unlimited food choices. The destructive voracity of the modern/colonial approach to taste normalized by gastronomy has made it hard to move toward more fair and sustainable food systems.

To transcend the coloniality of taste, we need to roll back the racialization, bureaucratization and desensualization of taste. We must develop an approach to the experience of taste that is not invested in the continuation of racialized modern/colonial distinctions and that does not prioritize market needs over human ones. This approach would be both more ethical and more satisfying than what the

DOI: 10.4324/9781003331834-7

normalized gastronomic conceptualization of taste has offered. While the decolonization of the global food system is an urgent and fundamental task, reclaiming the affectivity of taste suppressed by modernity and learning to cultivate the sense of taste as something other than an expression of a supposedly superior modern subjectivity is not a frivolous undertaking. No major challenge to the exploitative practices that sustain the global food system can be successful without delinking from the modern notion of taste, which drives and legitimizes the outsized consumption of food and resources. Learning from existing approaches to taste that do not depend on the otherization of affective taste, or on a constant supply of resources extracted from all over the world, would allow for less harmful forms of gustatory enjoyment to develop. The last section of this chapter discusses the kinds of movements and practices that evidence a yearning for more affective approaches to taste that could open paths toward the decolonization of the experience of taste.

The Coloniality of Taste

Today the foodie has taken the place of the gourmand as the main subject of the gastronomic discipline of taste. Foodies are likely to be defined as less classist and racist than gourmands, but the difference is not as meaningful as it might seem.[1] Foodies follow the logic of neoliberal multicultural inclusion that does not fundamentally change the unequal power relations of the modern/colonial world order. The gastronomic approach to taste inherited by foodies depends on and continues to facilitate the constantly expanding exploitation of peoples and landscapes that sustains the capitalist global food system. In the nineteenth century, French and British gastronomers bragged that the whole world toiled to supply their tables. Foodies today can be found in many more countries and, as privileged beneficiaries of the global food system, they command more than their fair share of the world's resources.

The coloniality of taste is not just a leftover from the nineteenth century. The processes of desensualization, bureaucratization and racialization of taste propelled by gastronomy have taken new forms. New technologies have multiplied the mediations between bodies and food. The desensualization of taste has been potentiated by audiovisual technologies, while the bureaucratization of taste is now advanced by food industries, thanks to food and flavor technologies. These mediations are ostensibly made in the name of pleasure, but they curtail the experience of taste to favor market imperatives. The racialization of taste, now obscured by the insufficient discourse of multicultural inclusion, fuels the relentless capitalist commodification that exploits nature, cultures and peoples. The racialization of the affective aspects of taste continues to mediate how taste is understood and experienced.

Desensualization

The gustatory experience has been increasingly audiovisualized: evoked through visual and auditory perception. Television and the Internet have an ever-growing offer of food-related content. Like gastronomic writing, food shows are proposed as providers of authoritative expertise to guide the consumption of food and the experience of taste, but most food shows aim primarily to entertain and to advertise products. As an audiovisual technology, television cannot engage the senses of taste and smell that define the gustatory experience. Instead, they provide food enjoyment through the eyes and ears. Food can be enjoyed as an audiovisual pleasure that does not involve gustation. While the visual aesthetization of food has expanded the ways in which food can be appreciated, it does not do much for the cultivation of the senses of taste and smell. As Steven Poole put it in his critique of gastroculture, "for a cooking show we have to outsource our sensory apparatus and with it our faculty of judgement."[2] Rather than enhancing the experience of gustatory taste as such, food television turns cooking and tasting into spectator sports.

The Internet has given additional force to the spectacularization of food. Internet applications like *Instagram* are used to circulate food images, and restaurant patrons are often more eager to photograph food than to actually appreciate its flavors. This has put increased pressure on restaurants to make the visual aspect of food a top priority. Applications like *Tasty* provide recipes in a video format that puts more emphasis on visual pleasure than on practicality. The images of food as it is being prepared show only the hands of the cook and there is background music instead of an instructional narrative. Ingredient quantities are written on the margins as they are added. *Tasty* provides a full recipe for those actually interested in cooking, but its recipes are designed to circulate on social media as a show of food effortlessly cooking itself.

Food media could in principle be informative and empower people to cook and explore taste, and in many cases it does. The Internet, in particular, has enabled the questioning of the bureaucratization of taste by exponentially expanding the number and variety of content producers. However, the cooking prowess and discriminating palate projected by food media celebrities can also disempower audiences by turning the basic daily activities of cooking, eating and even tasting into something regular people do not feel qualified to do properly on their own. The alienation from our ability to feed and sensorially satisfy ourselves benefits the food industry which, not surprisingly, is a major funder of food media. The audiovisual approach to taste has extended people's capacity to consume food well beyond the ingestion limitations of their bodies. When watching food media, people are consuming images, and they are offered the additional consumption of food products, cooking equipment, food travel and other commodities.

Bureaucratization

Food media plays a big role in the bureaucratization of taste in which experts and technologies mediate and limit the embodied experience of taste. Nineteenth-century gastronomy was characterized by the use of the printed word as a way of standardizing cooking and tasting. To achieve the objectivity of taste desired by gastronomers, cookbooks were used to establish the single best way of preparing each dish. Gastronomic writing was used to train eaters to respond to such preparations as if they were objectively the best ones. New technologies have amplified the bureaucratizing drive of gastronomy. Food and flavor technologies have joined the printed word and audiovisual technologies as powerful mediations that guide the experience of taste. Their impact is complex. Food and flavor technologies are used to develop products that appeal to and engage the senses, but they do it in a way that furthers the alienation from the experience of gustation.

The current era of consumer capitalism has been lauded as one in which pleasure has been democratized.[3] Most notably David Howes, director of the Concordia Sensoria Research Team in Montreal, has argued that the sensual characteristics of commodities have produced a state of constant sensory stimulation or "hyperesthesia."[4] Nevertheless, we should not forget that the global capitalist order attacks the senses of most humans on the planet. The exploitative conditions of labor and the sensual misery of the growing populations who live in shantytowns or in refugee camps are hard to imagine. Because of pollution and climate change, the attacks on the senses formerly endured mostly by industrial factory workers is now something everyone must endure in the world's megacities. Hyperesthesia, as the constant pleasant stimulation of the senses by commodities, is experienced only by a relatively small segment of the world's population. Even in that case, I would argue that hyperesthesia in no way represents an emancipation of the senses from capitalist alienation. The manipulation of the senses to spur the circulation of commodities only adds to the list of ways in which capitalism dehumanizes the senses, co-opting their capabilities. In consumer capitalism the senses are gratified but primarily as a means for consumption. As Marx put it, the entrepreneur tells customers: "I shall swindle you while providing you enjoyment."[5] The senses in consumer capitalism have been turned into an interface between human bodies and the market.

Gastronomic writers fantasized about food and taste technologies that would enhance the pleasures of taste. We still do not have technologies capable of enhancing, replicating or superseding the senses of smell and taste. What we do have is a multibillion-dollar flavor and fragrance industry that conducts sensory evaluation research to develop flavors that are added to food products. Flavor science and technologies have undeniable potential to contribute to the enhancement of sensory enjoyment. However, they have been developed primarily by and for the food industries. The industries' profit imperative has steered flavor research in a direction that is far from having fulfilled any utopia of sensory plenitude.

Technologies of sensory evaluation have tackled one of the main problems that concerned gastronomy: how to overcome the subjectivity of taste. Gastronomers saw the subjectivity of taste as incompatible with their aim to establish gastronomy as a science that can reach universally valid judgments, while the food industry sees the subjectivity of taste as limiting the profitability of its products. But while gastronomic writers attempted to suppress the subjectivity of taste, the developers of technologies of sensory evaluation have sought to understand it in order to influence it. Steven Shapin talks about technologies of sensory evaluation as an example of what he calls the "sciences of subjectivity," which aim to produce objective knowledge about subjective tastes.[6] The techniques that they have produced to achieve intersubjective judgments include the Flavor Profile, the Hedonic Index and focus groups. These techniques, Shapin argues, are world making since they shape our alimentary environment and everything else that is commercially formulated.[7]

Sensory evaluation techniques help to produce commercially successful products that appeal to widely shared taste preferences. The objective standards established by these techniques are intersubjective, but they are not universal. Flavors that are less widely appreciated, and the people who enjoy them, are excluded. If gastronomers prescribed the foods that they judged to be good, today the food industry is increasingly deciding which foods exist at all. The bureaucratization of taste is becoming more seriously consequential. Food industry research and decision making are usually kept in secrecy, because of the proprietary character of many of their methods and products. Big food industries increasingly have the capacity to decide which crops are grown globally and which food commodities are produced. If industrial agriculture threatens biodiversity by producing only a handful of crops, sensory evaluation techniques as currently employed threaten the diversity of the flavorscape. For example, the "Spectrum Descriptive Analysis," an influential sensory evaluation methodology that is used as a universal arbiter to evaluate the sensory properties of foods, is based on one dozen products from the U.S. food industry.[8] A food industry that only takes into account the sensory properties of a dozen of its own products, like Land O'Lakes American Cheese and Heinz Tomato Ketchup, is excluding an exorbitant number of organoleptic qualities from the foods that are available to consumers. It is reducing the richness of the flavorscape and shaping it in its own image.

Flavor chemistry analyzes flavor as compounds and molecules. Christy Spackman has argued that the gas chromatograph, in collaboration with expert human noses, has facilitated efforts to manage what flavor molecules are available to consumers.[9] However, flavor is a human response and there is no such thing as an electronic nose or tongue.[10] In flavor-related food science, the human nose is the most sensitive detection tool available.[11] The food industry has been able to design pleasing products, but not all products are commercially successful and they are never as universally successful as the industry would like. The sensitivity of the human nose has been a real challenge for the standardization of taste that the food industry would like to see happen. In an ethnographic study of food scientists in the United States, Ella Butler observed that scientists do not view consumers as easy to capture by the flavors of their products. Instead, they see the consumer's palate, and particularly its sensitivity to off flavors, as an obstacle.[12] From the perspective of the food and flavor industries, the superior sensory capabilities of the human body are a problem that needs to be controlled.

In response to the problems posed by the complexity of the human chemosenses (taste and smell), the food industry continues its efforts to understand the body's sensory responsiveness. Food scientists recognize the affective power of taste and smell, but they are channeling this power toward the goal of encouraging continuous consumption. One simple way of doing this has been the design of flavors that please but do not satisfy. An early market research discovery was that leading products end with a clean note, with little or no after taste, while the taste of second-place products lingers.[13] Flavors that cleanly disappear prompt consumers to take another sip or bite to repeat the flavor experience. This results in increased consumption, so from a market perspective the flavor is successful.[14] However, from a non-market perspective, such flavors could be seen as unsuccessful, since they keep consumers unsatisfied and encourage overconsumption that has negative sustainability and health consequences.

Contemporary chemosensory research is giving the food industry ever more potent tools to affect bodies and influence consumer behavior. Food industry research could offer valuable knowledge for the understanding of how smell and taste mediate human health. Nevertheless, as Sarah E. Tracy succinctly put it, the most authoritative knowledge of that sort is shaped by the corporate imperative to determine what chemical compounds humans register as pleasurable and thus what products they are likely to buy.[15] For the food industry, the finding that dietary intake of glutamate can induce flavor preference learning in rats has been more important than the need to understand how its reception in the gut affects the recognition of food intake, food digestion and absorption, and homeostatic processes.[16] Chemosensory research has empowered the food industry to produce crave-inducing foods whose potentially extensive ill effects are not well understood. The food industry no doubt recognizes the affectivity of taste, but it seeks to manage and control it for its own purposes.

Modern cooks have always aspired to be recognized as scientists, so it is not surprising that newer cooking trends have been inspired by the molecular approach to taste favored by food scientists. But whereas food scientists are striving to understand and manage the subjectivity and affectivity of taste for commercial purposes, cooks continue to be concerned with establishing the aesthetic value and objectivity of taste. Many celebrated contemporary chefs practice styles of cooking dubbed "modernist" and "molecular." Regardless of the suitability of the labels, these cooking styles are characterized by artistic exploration of the use of food additives and techniques developed by the food industry. "Modernist" chefs advocate for the wider availability of industrial additives and techniques, since they consider them as advancements that must be democratized. Ferrán Adriá, one of the most accomplished and vocal of these chefs, has developed a distinctive style of cooking characterized by its striking visual appeal. Adriá's work has been hailed by Jean-Paul Jouary as finally transforming cooking into an art in the "proper" aesthetic sense by fulfilling the requirements of originality, universality, representation and extension of understanding.[17] Adriá transformed cooking into an edible visual art, but his culinary creations tend to be more suitable for a museum than for the dinner table. Instead of validating cooking as a gustatory art by focusing on the aesthetic capabilities specific to the senses of taste and smell, high-end chefs continue to look up to the visual arts for validation.

Other contemporary trends in professional cooking have embraced new scientific knowledge and technologies in an effort to establish the definitive best way to cook everything. The desire to establish the best single way to prepare any given food has been a constant in gastronomic thought. The subjectivity of taste is disregarded for the sake of establishing cooking as an objective science. Scientists recognize the contingency of their objectivity, but in the field of cooking a naïve and authoritarian belief in universal objectivity prevails. A comprehensive example of this trend is the encyclopedic *Modernist Cuisine*. The authors of this five-volume set promise an up-to-date scientific take on cooking techniques and aim to bust myths regarding cooking and taste.[18] They promise the best procedure for cooking standard and innovative recipes using regular and industrial ingredients and equipment. The claim that their procedures are scientifically proven to be the best invokes science to overrule subjective taste differences. They downgrade the value of more experiential ways of cooking and transform the vital activity of food preparation into something too complicated for most people to attempt.

In principle there is no problem with the scientific exploration of food and flavor. The problem comes when the human experience of taste is subordinated to a form of scientific knowledge that has a poor understanding of how humans create and appreciate flavor and is more interested in serving market imperatives than anything else. Settling for the poor understanding of taste scientifically available and submitting to the imperatives of the market is impoverishing our sense of taste. Ironically, contemporary foodie culture underwrites the flattening of taste for the sake of having a constant supply of new food products. Foodism is the taste culture of consumer capitalism, insofar as it is driven by the desire for a constantly expanding range of food consumption possibilities. In the United States, food industries introduce over 20,000 new food products every year.[19] Most of these products are forgettable and disappear almost as fast as they come. But for foodies the most desired "taste" is novelty.[20] The consumerism of foodie culture exemplifies Jonathan Beller's idea that the alienation of the senses now takes the form of the sensuality of alienation.[21] According to Beller, we are in a "new modality of capitalism in which we participate culturally and socially vis-à-vis the senses in the production of our own impoverishment and liquidation."[22] In the case of modern food culture, the pleasures of food and food images keep us from grasping how our consumption fuels exploitative and unsustainable food production practices that threaten our health and global food security. It also depends on and reinforces racialized inequalities.

Racialization

The racialization of taste continues to affect the understanding and experience of taste, even though there seems to be more awareness about it. Colonialism deployed a racialized understanding of gustatory differences, invoking the more affective cuisines and table manners of the colonized as evidence of their supposedly natural inferiority. The idea that affective taste was the defining line separating the civilized from the inferior races made a lasting impact in Western societies as well as on those that have been subject to Western colonialism and imperialism. The experience of taste in modern and modernizing societies is mediated by the racialized understanding of gustatory affect in many different ways. Avoiding or embracing affective taste is always susceptible to being read as conveying belonging to, or risking exclusion from, "civilized" society.

Few people today may agree with the idea that affective taste is a sign of racial inferiority, but many are still apprehensive of embracing the affectivity of taste, particularly in public. The following is only one of countless possible anecdotal examples of how people curb their gustatory enjoyment for fear of receiving a negative social evaluation. A university professor did not eat their shell and head on shrimp when dining with U.S. students in Spain because they thought that eating with fingers would have brought down their status as a professor in the eyes of the students. The professor regretted having to abstain from eating the *gambas a la plancha*, which would have involved getting their fingers dripping with garlic oil while removing the shells and sucking on the juices of the just severed shrimp heads, perhaps making a little noise. This would have been a shocking or even disgusting spectacle for many of the students from the United States, who had only seen headless shelled shrimp at table. But it could also have been an opportunity to teach them about a more embodied experience of taste without stigmatizing it.

The normativity of unaffective taste is evident on any given day at schools and offices during lunchtime. It is common for the children of immigrants in the United States to request "normal" food for their school lunches after their peers make fun of them for eating "smelly" food. The experience is repeated later in life when coworkers scold anybody who dares to bring food that according to them "smells up" the place. The bodies of those who have internalized the modern discipline of taste find olfactorily affective foods repulsive. They are unable to participate in food experiences with those who have not been subjected to, or have self-consciously rejected, the modern discipline of taste. But in societies in which the modern is understood as superior, those who have closed themselves to the affectivity of taste have the power of imposing unaffective taste on others, at least in public settings. I am sure readers can provide many similar examples of their own.

An early example of how colonized societies managed the modern requirement to curb the affectivity of taste can be found in the writings of nineteenth-century nation builders in Bengal, India. Cultural anthropologist Bhaskar Mukhopadhyay explains how nationalist reformers purged red chilies and raw tamarind from the polite Bengali palate as a part of a "civilizing" and westernizing process.[23] Influential nineteenth-century literary figure Bankimchandra Chattopadhyay decried tamarind: "To tell the truth, I cannot find anything under the sun as harmful as tamarind. Whoever eats it gets acidity and belches...."[24] Chattopadhyay's choice of tamarind and red chili as the representation of the gustatory and affective excess that has to be purged to enter the realm of the modern and civilized reflects an astute negotiation with the modern/colonial stigmatization of affective taste. Rather than renouncing the affectivity of taste as represented by spices in general like they did in Europe, he abjured only red chili and raw tamarind. According to Mukhopadhyay, these two banished flavors became a trademark of Bengali street foods, which have always been considered off limits for the well to do.

The colonial racialization of affective taste, as well as the nation builder's negotiation with it, continues to inform the way many contemporary middle-class Bengalis relate to affective taste. While discussing the nineteenth-century response to the racialization of taste, Mukhopadhyay confesses his own inability to enjoy chili and tamarind-based street foods. As a self-described middle-class and middle-aged Bengali, he explains that his body cannot tolerate these flavors because being raised drinking contaminated water made his stomach delicate and because it wouldn't be respectable anyway.[25] He reports watching a teenage girl eat a tamarind and chili laden street snack called *fuska* and being fascinated by her feat which he was convinced "must have set her whole alimentary system on fire."[26] Then he mused: "Yet she looked almost ravished with her burning tongue sticking out and pupils dilated; one could not possibly miss the sense of exhilaration writ large on her face, even as drops of tears descended her pimply cheeks."[27] Mukhopadhyay admires the unrestrained affective pleasure evident in the teenage girl's body, but he cannot enjoy his own *fuska* in the same way. Despite his critical understanding of the curbing of affective taste as a tool for policing colonial and social hierarchies, Mukhopadhyay cannot overcome his own body's internalization of the modern Bengali discipline of taste and therefore cannot enjoy foods that he knows are enjoyable.

Colonialism affected the food cultures of the colonized around the world in many ways, including the destruction of the material conditions of possibility of their food and taste cultures as well as their epistemic and aesthetic devaluation. As a way of contending with the devaluation of their culinary cultures, the elites of postcolonial nations and of countries that were otherwise subjected to Western imperialism have taken it upon themselves to "modernize" their cuisines. This has meant documenting and reforming their cuisines in ways that would be understandable and not disapproved of by the imperial Western gaze. The sensorial Eurocentrism instituted by gastronomic thought has put the peoples that Western modernity marginalized in a quandary. They either disavow the affectivity of taste to demonstrate their worthiness as modern and civilized, or they preserve it but continue to be categorized and treated as inferior. These days, foodie discourse values the "authenticity" of so-called ethnic cuisines very highly. Given the ongoing coloniality of global power relations, the demand for authenticity could be seen as a way of encouraging others to keep the culinary and sensorial traits that underwrite their continued subordination and exploitation.

In Turkey, the quandary posed by the racialization of taste became a national debate. Defne Karaosmanoğlu has explained how governmental and private culinary institutes, as well as chefs, gourmets and food writers, embarked on a process of "modernizing" Turkish cuisine in an effort to represent the country as not incompatible with the West.[28] This process can be understood as an exercise of gastrodiplomacy to aid Turkey's bid to join the European Union.[29] Joining the European Union seems to require a culinary culture that wouldn't clash with the modern discipline of taste. One of the critics of the efforts to modernize Turkish cuisine sees it as an act of submission to the West: "Turkish cuisine is modern already. What should be changed is the inferiority complex."[30] Those in favor of reforming Turkish cuisine want to remake it in the image of Europe hoping that lessening culinary and sensory differences would lead to political and economic equality.

The proposed new Turkish cuisine echoes the trajectory of the Nuevo Latino culinary trend of a few decades ago. Nuevo Latino's "modernization" of Latin American cuisines implied toning down affective flavors and increasing the visual appeal of the food. It also involved the application of French culinary techniques, which still define high-end professional cooking in the West.[31] The Nuevo Latino transformation of Latin American cuisines is predicated on the idea that only French-influenced cuisines have a valid sensory logic and culinary technique. It treats Latin America as a source of ingredients devoid of any valuable cultural elaboration. One of the latest configurations of the racialization of taste can be seen in the "gastronomic revolution" in Perú, as discussed by María Elena García.[32] There are many more examples from around the world, and a growing bibliography that analyzes them.

While modernizing sectors of many countries are modifying their taste cultures to avoid being marginalized in a global market dominated by Western standards, food media and food industries continue to mobilize racist tropes to attract viewers and customers. A racialized understanding of taste and geography permeates food travel shows. Casey Ryan Kelly has discussed how culinary adventure shows like *Bizarre Foods* presuppose the superiority of "Western" values and maintain colonial "distinctions between clean, orderly, and civilized eating rituals of the First World and the strange, primal and uncanny cuisines that define the 'Third World.'"[33] The framing of affective foods as weird or even "disgusting" dehumanizes the people who eat them, particularly if they belong to a group that is already marginalized. The dehumanization of others allows the viewers of culinary adventure shows to feel racially superior. It also adds a transgressive thrill to the viewer's enjoyment of commodified versions of the foods of otherized and marginalized peoples. Elements from cuisines from all over the world have been transformed into easy-to-consume standardized novelty products divorced from their cultural, environmental and culinary contexts, and which are far from having the same sensory qualities as the foods they purportedly replicate. For example *Trader Joe's*, with its kitschy reformulation of the colonial explorer trope, is a foodie paradise in which foods from around the world are available in ready-to-heat and eat versions. Foodie cosmopolitanism is colonialist consumerism.

The experience of taste that has been normalized as modern continues to be significantly limited by the processes of desensualization, bureaucratization and racialization that were propelled by nineteenth-century gastronomic writers. These processes have acquired a new force from food industry-led sensory research and food technologies, and from audiovisual food media. The idea that gastronomy liberated taste and that consumer capitalism has delivered an era of sensory plenitude obfuscates our understanding of the ways in which the modern approach to taste is harmful to those it gratifies as well as to those it exploits materially and culturally. However, signs of increasing dissatisfaction with the coloniality of taste are apparent, and decolonial paths are being opened.

Decolonial Taste Yearnings

Decolonial movements are as old as colonialism. Colonized, enslaved and marginalized peoples have always creatively struggled against the destruction and devaluation of their ways of producing, preparing and consuming food. Many food-related initiatives today are directly challenging the coloniality of the global food production system. A growing number of movements all over the world are working to create more sustainable and fair food systems. In the last 25 years, the transnational peasant movement called *La Vía Campesina* has worked toward the goal of food sovereignty. This goal involves "the right of peoples to healthy and culturally appropriate food produced through ecologically sound and sustainable methods, and their right to define their own food and agriculture systems."[34] This implies a rejection of the corporate food and trade regime and new social relations free from oppression and inequality.[35] Indigenous communities all over the world are also defending their right to sustainable self-determined development.[36] These and many other movements aspire to create a world in which different food economies and agricultural and culinary knowledges and practices can coexist in conditions of equality.

Food sovereignty movements reaffirm the validity of food production systems that have been attacked by capitalist modernity; they also enable the revitalization of their culinary practices. They are an example of what Adolfo Albán Achinte has called practices of re-existence. He defines re-existence as the re-elaboration of life in the attempt to overcome adverse conditions and inhabit a place of dignity in society.[37] Albán Achinte developed this concept in his account of the history of the culinary practices of Afro-descendant communities in Colombia and Ecuador,[38] but examples of culinary practices of re-existence can be found around the world. The reaffirmation of food systems and culinary cultures that were marginalized by modernity is a matter of physical survival as much as it is a matter of dignity. It is not a nativist defense of "authenticity" that presents people as static. The revitalization of culinary practices that were slated for extinction by capitalist modernity also involves sensory decolonization.

Signs of discontent with the unaffective approach to taste that gastronomy normalized as modern have always existed and they are not hard to find. There are thinkers, activists, food producers and food consumers all over the world who continue to practice, envision and forge decolonial taste options. These approaches to the experience of taste are more fulfilling and less harmful to self and others than the one offered by modernity/coloniality. Gastronomic writers built their notion of taste in an effort to overcome how modern epistemology and aesthetics had devalued this sense, renouncing the subjectivity and affectivity of taste in the process. But dissatisfaction with this gastronomic compromise has taken new force in contemporary Western philosophy. The push to validate the aesthetic and epistemic capabilities of taste is bringing Western philosophy closer to longstanding philosophical ideas prevalent in other parts of the world, where such capabilities have always been recognized to different degrees. A few contemporary philosophers in the Western tradition are using gustatory taste as a model to expand the limited ways in which art and knowledge have been understood in Western philosophy. Raymond D. Boisvert and Lisa Heldke, for example, consider that the concept of knowledge should be modeled after the sense of taste. Their concept of "knowing-as-tasting" conveys that knowing is an interactive process that happens in a given context and rejects the standard notion of "knowing-as-seeing" in which an objective knower is detached from the known.[39] Instead of putting down taste because it does not fit epistemic and aesthetic standards, recognizing the complexity of taste is allowing thinkers to question the validity of such standards. Their work is consistent with the work of thinkers and artists from different parts of the world who have been advocating for a decolonial aesthesis, which implies subverting the modern/colonial control over the senses and perception.[40]

Dissatisfaction with the modern culture of taste can also be seen in consumers of industrial food products and food media. In *Appetite for Change: How the Counterculture Took on the Food Industry*, Warren Belasco chronicled the rise of the alternative food movements in the United States. Unfortunately, as Belasco admits in the preface to the second edition of the book: "Yet, even with Thoreau and Gandhi on our side, we failed to change the world -or ourselves- very much."[41] It is sad but imperative to realize that the main achievement of alternative food movements was the creation of new spheres of production and consumption. Food products marketed as organic, vegetarian, low fat, sugar free, fair trade, "ethnic" and so on have done more to diversify and enrich the food industry than to advance the goals of the alternative food movements.

Equally disappointing is the outcome of "Slow Food," an international organization founded in 1986 in Italy. Championing the slow life and sensory pleasure, Slow Food has grown into a bureaucratic international organization. Whereas the goals of cultivating taste and preserving biodiversity are worthwhile, the organization's founder, Carlo Petrini, has not been able to overcome the logic of the market. He has argued that "in order to live pleasurably, we need to broaden the range of things that give us pleasure."[42] This frame of mind has sent slow food members all over the world looking for rare local foods in the name of preservation. There is a shocking lack of self-reflexivity in Petrini when he proudly claims that *Salone del Gusto*, Slow Food's annual central gathering, "has emerged as a new form of "cultural" marketing, a thousand times more effective than a trade fair, both from the commercial point of view and in terms of the promotion of our strategic goals."[43] While presenting itself as an alternative to fast food, the Slow Food movement does not interrogate its capitalist and colonialist sense of entitlement to the alimentary resources of the whole planet.

Alternative food consumption movements have not been effective in overcoming the coloniality of the food system, but they are testimony to a widespread discontent with the modern discipline of taste. The Internet, more than any new technology, has allowed for this discontent to be expressed. Food blogs multiplied the number of people who could circulate their perspectives on cooking and eating, which has eroded the authoritative claims of food publications and television shows. Different perspectives on cooking and taste have always existed, but they had been obscured by the claims to universal objectivity and superiority of the few who could publish books and magazines or produce and broadcast television shows. Now people on the Internet routinely question the authority of food-related content providers. In the ratings and comments section of *The New York Times Cooking* application, for example, readers write extensive reviews of the recipes. Many praise the recipes, but they also report on the many ways in which they changed them and why. The original recipe and the accumulated comments allow subsequent readers to decide which ingredients and procedures might work best for them, based on virtual collective experience and their own preferences. The authority of the objective and universal bureaucrat of taste is unraveling, making space for the articulation of different gustatory intersubjectivities.

A rather dramatic example of dissatisfaction with the modern discipline of taste can be found in the eating shows known as *mukbang*, which have origins in Japan but really took off in South Korea in 2009.[44] *Mukbang* are live streamed on the Internet, and they are produced by individuals who eat large quantities of food while talking to their viewers, reading their chat messages and receiving donations. The most famous *mukbang* hosts have hundreds of thousands of followers and make hundreds of thousands of dollars in earnings a year.[45] Versions of these shows are beginning to emerge in the United States, although they have not reached the same levels of success. Many elements of these shows can be attributed to specific aspects of contemporary South Korean society, but they are also clearly shaped by more widespread aspects of late capitalist consumer society and digitally mediated culture. Yeran Kim, after analyzing the many contradictory aspects of the practice of *Mukbang*, concluded that in it "the vivid expression of transgressive affect is exploited as a novel form of digital commodity."[46] The transgressive elements of the shows, even if they are co-opted by capitalism, demonstrate a yearning for a more affective interaction with food and taste. *Mukbang* fans get virtual dining company and the vicarious satisfaction of eating in ways that are not considered socially acceptable in their milieu. Mukbang hosts break the rules regarding the amount of food that is acceptable or even possible to eat, which is problematic in many ways, but they also break the rules regarding how to enjoy the experience of gustation. They histrionically break the rules of middle-class manners. *Mukbang* fans expect the hosts to describe their sensations while eating, produce eating and delight noises, and get their fingers and lips messy.[47] The hosts' exaggerated eating performance is appreciated as a rebellious gesture against the restrictions of the modern discipline of taste, which stifle embodied gustatory pleasure.

A more active disobedience of the modern discipline of taste, which racialized embodied and affective taste as a mark of inferiority, can be found in everyday practices all over the world. This disobedience is becoming more visible in two of the bastions of the gastronomic discipline: restaurants and food writing. Restaurants are the foundational disciplinary space where gastronomic subjects are tested. Restaurant patrons are expected to behave in ways that demonstrate their knowledge and adhesion to the modern discipline of taste. Likewise, writing was the tool of choice for the codification of gastronomic "laws." A couple of examples will illustrate how restaurants and food writing are being repurposed as tools for gustatory decolonization.

Eating with fingers, which allows for the tactile enjoyment of food before it is placed in the mouth, has been one of the main sensory pleasures of eating that have been censored by the modern discipline of taste. Gastronomic writers created a racialized hierarchy of civilizations based on the presence or absence of eating utensils, the kind of utensils, and the ways of employing them. They established eating with knife and fork as the highest expression of civilization. Given the normalization of the idea that eating with fingers is primitive, it is remarkable that one of the most elite restaurants in India does not provide eating utensils. The restaurant *Bukhara* in New Delhi has been visited by world leaders and celebrities. It offers cutlery on request, but the expected norm is to eat with fingers; cutlery is provided only as an accommodation. Restaurant patrons confidently and skillfully eat with their fingers in the luxurious setting in explicit repudiation of the modern discipline of taste. This is not to say that manners are not involved. Eating with fingers of course has its own techniques and etiquette that allow for social distinctions. But the absence of utensils is not in itself a mark of low status. Eating with fingers is just a part of the sensory appreciation and enjoyment of food. Other upscale restaurants in India offer cutlery by default but eating with fingers is always welcome. It is not uncommon for restaurant patrons to invite non-Indian visitors to try to eat with fingers to experience their food more fully.

Confident sensory disobedience can also be found explicitly in gastronomic writing. Vertamae Smart-Grosvenor shared in her memoirs her regrets over having adapted to eating everything with cutlery while living in Europe:

> being the granddaughter of a slave who adapted to the unnatural ways of his master, I, too, soon caught on and there I was eating fruit with a fork. How unnatural can you get! Big juicy orange and you got to take it in fork-fuls instead of letting all that juice run on your hands and then licking your fingers.[48]

Smart-Grosvenor understood that the pressure to adapt to the ways of the powerful deprived her of sensory pleasures, and eventually went back to eating fruit with fingers. It takes decolonial confidence to reclaim sensory pleasures when others can use that to rob you of human dignity.

Another example of sensory disobedience can be found in the work of Nigerian food writer Yemisí Aríbisálà, who is keenly aware that Nigerian food has not been eulogized in writing as much as other cuisines and begins to make up for that in her memoirs.[49] Her account of the wide variety of cuisines, dishes and condiments of her country is full of admiration and contains vivid descriptions of how Nigerian dishes maximize the organoleptic and affective qualities of food. While living in England, Aríbisálà befriended people from different countries who would tell her things like that the Nigerian food of a neighbor "smelled like shit" and who were convinced that Nigerian food was "jungle fare" not worth talking about.[50]

Writing about Nigerian food in a praiseful way is transgressive in itself. But Aríbisálà also challenges the conventions of food writing by celebrating food qualities that the modern discipline of taste has dismissed as undesirable like mucilaginous textures, the sensation of heat and pungent aromas. The mucilaginous texture that is widely disparaged in Western culinary cultures is the defining quality of a whole category of soups in Nigeria known as *draw* soups. Aríbisálà explains that *draw* "refers to that thing that connects two kissers moving apart to catch their breath: organic strings pulling in two directions."[51] Her description of *draw* conveys the almost erotic pleasure of mucilaginous textures like the one produced by okra.

Aríbisálà's description of consuming a chili hot dish as a full body experience is equally evocative. She explains that when eating peppered dishes called *yíláta* at a popular mall in Lagos

> your nose ran, your eyes watered, you got hot under the collar and every-where else. You wept and fanned yourself in vain from the effects of the quantities of pepper applied to the yíláta. You laughed at your own stupidity. You drank glasses of water and your temperature soared dangerously. But underneath the pepper, the snails or chicken or ram was so delicious, you couldn't help continuing to eat and suffer.[52]

Aríbisálà also praises the flavor-boosting properties of dried codfish head:

> This head has sailed all the way from Norway, from under the turned up noses of those silly Europeans who don't understand the first thing about flavour. Give thanks for it and for their snobbery towards many things wonderfully pungent and malodorous and ugly.[53]

From Aríbisálà's non-Eurocentric perspective, the modern discipline of taste is revealed as sensorially and affectively impoverished.

Smart-Grosvenor and Aríbisálà have put in print decolonial thoughts shared by peoples around the world who refuse to be shamed into limiting their sensory experience and gustatory pleasure by racist modern/colonial discourses. Foodies and those who are on the privileged side of the modern/colonial regime have much to learn from the defense of the embodied and affective experience of taste. It is time to overcome the gastronomic mystique and recognize that the modern/colonial discipline of taste is harmful to us all.

Notes

1 For a discussion of elitism in foodie discourse, see Josée Johnston and Shyon Baumann, *Foodies: Democracy and Distinction in the Gourmet Foodscape* (London: Routledge, 2009).
2 Steven Poole, *You Aren't What You Eat: Fed up with Gastroculture* (Plattsburgh, NY: McClelland & Stewart, 2012), 162.

3 Luca Vercelloni, "The Economy of Taste in Consumer Society," in *The Invention of Taste* (London: Bloomsbury Publishing Plc, 2016), 95–158, https://doi.org/10.5040/9781474273633.ch-005.

4 David Howes, "Hyperesthesia, or, The Sensual Logic of Late Capitalism," in *Empire of the Senses: The Sensual Culture Reader*, ed. David Howes (Oxford; New York: Berg, 2005), 281–303.

5 Karl Marx, *Karl Marx Early Writings*, trans. T.B. Bottomore (New York; Toronto; London: McGraw-Hill, 1964), 169.

6 Steven Shapin, "The Sciences of Subjectivity," *Social Studies of Science* 42, no. 2 (April 2012): 170–184, https://doi.org/10.1177/0306312711435375.

7 Shapin, 179.

8 Jacob Lahne, "Standard Sensations: The Production of Objective Experience from Industrial Technique," *The Senses and Society* 13, no. 1 (January 2, 2018): 6–18, https://doi.org/10.1080/17458927.2017.1420842.

9 Christy Spackman, "Perfumer, Chemist, Machine: Gas Chromatography and the Industrial Search to 'Improve' Flavor," *The Senses and Society* 13, no. 1 (January 2, 2018): 54, https://doi.org/10.1080/17458927.2018.1425210.

10 Ella Butler, "Tasting Off-Flavors: Food Science, Sensory Knowledge and the Consumer Sensorium," *The Senses and Society* 13, no. 1 (January 2, 2018): 80, https://doi.org/10.1080/17458927.2017.1420028.

11 Butler, 81.

12 Butler, 79.

13 Nadia Berenstein, "Designing Flavors for Mass Consumption," *The Senses and Society* 13, no. 1 (January 2, 2018): 32, https://doi.org/10.1080/17458927.2018.1426249.

14 Berenstein, 32.

15 Sarah E. Tracy, "Delicious Molecules: Big Food Science, the Chemosenses, and Umami," *The Senses and Society* 13, no. 1 (January 2, 2018): 89, https://doi.org/10.1080/17458927.2017.1420027.

16 Tracy, 99–100.

17 Jean-Paul Jouary and Ferrán Adriá, *Ferrán Adrià and El Bulli: The Art, the Philosophy, the Gastronomy* (New York: Overlook Press, 2014).

18 Nathan Myhrvold et al., *Modernist Cuisine: The Art and Science of Cooking* (Bellevue, WA: Cooking Lab, 2011).

19 "USDA ERS - New Products," accessed June 22, 2021, https://www.ers.usda.gov/topics/food-markets-prices/processing-marketing/new-products/.

20 For a discussion of how novelty has been a constant in gastronomic thought see Alberto Capatti, *Le Goût du nouveau: Origines de la modernité alimentaire* (Paris: A. Michel, 1989).

21 Jonathan Beller, *The Cinematic Mode of Production: Attention Economy and the Society of the Spectacle* (Hanover, NH: Dartmouth College Press, 2006), 245.

22 Beller, 250.

23 Bhaskar Mukhopadhyay, "Between Elite Hysteria and Subaltern Carnivalesque: The Politics of Street-Food in the City of Calcutta," *South Asia Research* 24, no. 1 (May 2004): 37–50, https://doi.org/10.1177/0262728004042762.

24 Quoted in Mukhopadhyay, p. 43.

25 Mukhopadhyay, 37–38.

26 A fuska is a small hollow ball of fried dough, stuffed with spiced lentils, potatoes and spicy tamarind water, 37–38.

27 Mukhopadhyay, 38.

28 Defne Karaosmanoglu, "Surviving the Global Market: Turkish Cuisine 'Under Construction,'" *Food, Culture and Society: An International Journal of Multidisciplinary Research* 10, no. 3 (Fall 2007): 425–448.

29 Karaosmanoglu, 440.

30 Quoted in Karaosmanoglu, 433.

31 For a discussion of this see Zilkia Janer, "(In)Edible Nature: New World Food and Coloniality," *Cultural Studies* 21, no. 2 (March 2007): 385–405. https://doi.org/10.1080/09502380601162597.

32 María Elena García, *Gastropolitics and the Specter of Race: Stories of Capital, Culture, and Coloniality in Peru* (Oakland, CA: University of California Press, 2021).

33 Casey Ryan Kelly, *Food Television and Otherness in the Age of Globalization* (Lanham; Boulder; New York; London: Lexington Books, 2017), 11.

34 Via Campesina, "Twenty-Five Years of Envisioning Food Sovereignty: Celebrating Diversity, Resilience, and Transforming the Society," https://viacampesina.org/en/twenty-five-years-of-envisioning-food-sovereignty-celebrating-diversity-resilience-and-transforming-the-society/. Accessed June 2022.

35 Via Campesina

36 Tebtebba Foundation, *Sustaining and Enhancing Indigenous Peoples' Self-Determined Development: 20 Years After Rio*, ed. Joji Cariño et al. (Baguio City, Philippines: Tebtebba Foundation, 2012).

37 Adolfo Albán Achinte, *Sabor, poder y saber: Comida y tiempo en los valles afroandinos del Patía y Chota-Mira* (Popayán, Colombia: Editorial Universidad del Cauca, 2015), Note 27 pp. 39–40.

38 Albán Achinte.

39 Raymond D. Boisvert and Lisa M. Heldke, *Philosophers at Table: On Food and Being Human* (London: Reaktion Books, 2016), loc. 1833 of 3399, Kindle.

40 Walter D. Mignolo and Rolando Vázquez, "Decolonial AestheSis: Colonial Wounds/Decolonial Healings," *Social Text*, July 15, 2013, https://socialtextjournal.org/periscope_article/decolonial-aesthesis-colonial-woundsdecolonial-healings/. Accessed June, 2022.

41 Warren J. Belasco, *Appetite for Change: How the Counterculture Took on the Food Industry*, 2nd updated ed. (Ithaca, NY: Cornell University Press, 2007), 10.

42 Carlo Petrini, *Slow Food: The Case for Taste* (New York: Columbia University Press, 2003), 21.

43 Petrini, 63.

44 Yeran Kim, "Eating as a Transgression: Multisensorial Performativity in the Carnal Videos of *Mukbang* (Eating Shows)," *International Journal of Cultural Studies* 24, no. 1 (January 2021): 110, https://doi.org/10.1177/1367877920903435.

45 "MUNCHIES Presents: Mukbang," https://www.vice.com/en/article/vvx53d/munchies-presents-mukbang. Accessed June, 2022.

46 Kim, 119.

47 "MUNCHIES Presents."

48 Vertamae Smart-Grosvenor, *Vibration Cooking or The Travel Notes of a Geechee Girl* (New York: Ballantine Books, 1992), 72–73.

49 Yemisí Aríbisálà, *Longthroat Memoirs: Soups, Sex and Nigerian Taste Buds* (Abuja; London: Cassava Republic Press).

50 Aríbisálà, 214.

51 Aríbisálà, 149.

52 Aríbisálà, 282.

53 Aríbisálà, 222–223.

References

Albán Achinte, Adolfo. *Sabor, poder y saber: Comida y tiempo en los valles afroandinos del Patía y Chota-Mira*. Popayán, Colombia: Editorial Universidad del Cauca, 2015.

Aríbisálà, Yemisí. *Longthroat Memoirs: Soups, Sex and Nigerian Taste Buds*. Abuja; London: Cassava Republic Press, 2016.

Belasco, Warren James. *Appetite for Change: How the Counterculture Took on the Food Industry*. 2nd updated ed. Ithaca, NY: Cornell University Press, 2007.

Beller, Jonathan. *The Cinematic Mode of Production: Attention Economy and the Society of the Spectacle*. Hanover, NH: Dartmouth College Press, 2006.

Berenstein, Nadia. "Designing Flavors for Mass Consumption." *The Senses and Society* 13, no. 1 (January 2, 2018): 19–40. https://doi.org/10.1080/17458927.2018.1426249.

Boisvert, Raymond D., and Lisa M. Heldke. *Philosophers at Table: On Food and Being Human*. London: Reaktion Books, 2016.

Butler, Ella. "Tasting Off-Flavors: Food Science, Sensory Knowledge and the Consumer Sensorium." *The Senses and Society* 13, no. 1 (January 2, 2018): 75–88. https://doi.org/10.1080/17458927.2017.1420028.

Capatti, Alberto. *Le Goût du nouveau: Origines de la modernité alimentaire*. Paris: A. Michel, 1989.

García, María Elena. *Gastropolitics and the Specter of Race: Stories of Capital, Culture, and Coloniality in Peru*. Oakland, CA: University of California Press, 2021.

Howes, David. "Hyperesthesia, or, The Sensual Logic of Late Capitalism." In *Empire of the Senses: The Sensual Culture Reader*. Edited by David Howes, 281–303. Oxford; New York: Berg, 2005.

Janer, Zilkia. "(In)Edible Nature: New World Food and Coloniality." *Cultural Studies* 21, no. 2 (March 2007): 385–405. https://doi.org/10.1080/09502380601162597.

Johnston, Josée, and Shyon Baumann. *Foodies: Democracy and Distinction in the Gourmet Foodscape*. London: Routledge, 2009.

Jouary, Jean-Paul, and Ferrán Adriá. *Ferrán Adriá and El Bulli: The Art, the Philosophy, the Gastronomy*. New York: Overlook Press, 2014.

Karaosmanoglu, Defne. "Surviving the Global Market: Turkish Cuisine 'Under Construction.'" *Food, Culture and Society: An International Journal of Multidisciplinary Research* 10, no. 3 (Fall 2007): 425–448.

Kelly, Casey Ryan. *Food Television and Otherness in the Age of Globalization*. Lanham; Boulder; New York; London: Lexington Books, 2017.

Kim, Yeran. "Eating as a Transgression: Multisensorial Performativity in the Carnal Videos of *Mukbang* (Eating Shows)." *International Journal of Cultural Studies* 24, no. 1 (January 2021): 107–122. https://doi.org/10.1177/1367877920903435.

Lahne, Jacob. "Standard Sensations: The Production of Objective Experience from Industrial Technique." *The Senses and Society* 13, no. 1 (January 2, 2018): 6–18. https://doi.org/10.1080/17458927.2017.1420842.

Marx, Karl. *Karl Marx Early Writings*. Translated by T.B. Bottomore. New York; Toronto; London: McGraw-Hill, 1964.

Mignolo, Walter D., and Rolando Vázquez, "Decolonial AestheSis: Colonial Wounds/Decolonial Healings," *Social Text*, July 15, 2013, https://socialtextjournal.org/periscope_article/decolonial-aesthesis-colonial-woundsdecolonial-healings/. Accessed June, 2022.

Mignolo, Walter, and Catherine E. Walsh. *On Decoloniality: Concepts, Analytics, Praxis*. Durham, NC: Duke University Press, 2018.

Mukhopadhyay, Bhaskar. "Between Elite Hysteria and Subaltern Carnivalesque: The Politics of Street-Food in the City of Calcutta." *South Asia Research* 24, no. 1 (May 2004): 37–50. https://doi.org/10.1177/0262728004042762.

"MUNCHIES Presents: Mukbang." Accessed June, 2022. https://www.vice.com/en/article/vvx53d/munchies-presents-mukbang.

Myhrvold, Nathan, Chris Young, Maxime Bilet, and Ryan Matthew Smith. *Modernist Cuisine: The Art and Science of Cooking*. Bellevue, WA: Cooking Lab, 2011.

Petrini, Carlo. *Slow Food: The Case for Taste*. New York: Columbia University Press, 2003.

Poole, Steven. *You Aren't What You Eat: Fed up with Gastroculture*. Plattsburgh, NY: McClelland & Stewart, 2012.

Shapin, Steven. "The Sciences of Subjectivity." *Social Studies of Science* 42, no. 2 (April 2012): 170–184. https://doi.org/10.1177/0306312711435375.

Smart-Grosvenor, Vertamae. *Vibration Cooking or The Travel Notes of a Geechee Girl*. New York: Ballantine Books, 1992.

Spackman, Christy. "Perfumer, Chemist, Machine: Gas Chromatography and the Industrial Search to 'Improve' Flavor." *The Senses and Society* 13, no. 1 (January 2, 2018): 41–59. https://doi.org/10.1080/17458927.2018.1425210.

Tebtebba Foundation. *Sustaining and Enhancing Indigenous Peoples' Self-Determined Development: 20 Years After Rio*. Edited by Joji Cariño, Raymond de Chavez Wessendorf, Tirso Gonzales, and Ma. Elena Regpala. Baguio City, Philippines: Tebtebba Foundation, 2012.

Tracy, Sarah E. "Delicious Molecules: Big Food Science, the Chemosenses, and Umami." *The Senses and Society* 13, no. 1 (January 2, 2018): 89–107. https://doi.org/10.1080/17458927.2017.1420027.

"USDA ERS - New Products." https://www.ers.usda.gov/topics/food-markets-prices/processing-marketing/new-products/. Accessed June, 2022.

Via Campesina. "Twenty-Five Years of Envisioning Food Sovereignty: Celebrating Diversity, Resilience, and Transforming the Society : Via Campesina," June 28, 2021. https://viacampesina.org/en/twenty-five-years-of-envisioning-food-sovereignty-celebrating-diversity-resilience-and-transforming-the-society/. Accessed June, 2022.

Vercelloni, Luca. "The Economy of Taste in Consumer Society." In *The Invention of Taste*, 95–158. London: Bloomsbury Publishing Plc, 2016. https://doi.org/10.5040/9781474273633.ch-005.

INDEX

Pages followed by n refer notes.

Printed in the United States
by Baker & Taylor Publisher Services